W9-CRB-790

ACS SYMPOSIUM SERIES **490**

Flavor Precursors

Thermal and Enzymatic Conversions

Roy Teranishi, EDITOR
Agricultural Research Service
U.S. Department of Agriculture

Gary R. Takeoka, EDITOR
Agricultural Research Service
U.S. Department of Agriculture

Matthias Güntert, EDITOR
Haarmann and Reimer GmbH

Developed from a symposium sponsored
by the Division of Agricultural and Food Chemistry
at the Fourth Chemical Congress of North America
(202nd National Meeting
of the American Chemical Society),
New York, New York,
August 25–30, 1991

American Chemical Society, Washington, DC 1992

SEP/AE
CHEM
04789453

Library of Congress Cataloging-in-Publication Data

American Chemical Society. Meeting (202nd: 1991: New York, N.Y.)
  Flavor precursors: Thermal and enzymatic conversions / Roy Teranishi, Gary R. Takeoka, Matthias Güntert; developed from a symposium sponsored by the Division of Agricultural and Food Chemistry at the Fourth Chemical Congress of North America (202nd National Meeting of the American Chemical Society), New York, New York, August 25–30, 1991.

  p.   cm.—(ACS symposium series; ISSN 0097–6156; 490)

  Includes bibliographical references and index.

  ISBN 0–8412–2222–3

  1. Flavor—Congresses.   2. Biochemistry—Congresses.   3. Food—Composition—Congresses. 4. Metabolism—Congresses.

  I. Teranishi, Roy, 1922–   . II. Takeoka, Gary R. III. Güntert, Matthias. IV. American Chemical Society. Division of Agricultural and Food Chemistry. V. Chemical Congress of North America (4th: 1991: New York, N.Y.) VI. Title. VII. Series.

QP801.F45A43   1992
664′.07—dc20   92–10454
                                                              CIP

The paper used in this publication meets the minimum requirements of American National Standard for Information Sciences—Permanence of Paper for Printed Library Materials, ANSI Z39.48–1984.   ∞

Copyright © 1992

American Chemical Society

All Rights Reserved. The appearance of the code at the bottom of the first page of each chapter in this volume indicates the copyright owner's consent that reprographic copies of the chapter may be made for personal or internal use or for the personal or internal use of specific clients. This consent is given on the condition, however, that the copier pay the stated per-copy fee through the Copyright Clearance Center, Inc., 27 Congress Street, Salem, MA 01970, for copying beyond that permitted by Sections 107 or 108 of the U.S. Copyright Law. This consent does not extend to copying or transmission by any means—graphic or electronic—for any other purpose, such as for general distribution, for advertising or promotional purposes, for creating a new collective work, for resale, or for information storage and retrieval systems. The copying fee for each chapter is indicated in the code at the bottom of the first page of the chapter.

The citation of trade names and/or names of manufacturers in this publication is not to be construed as an endorsement or as approval by ACS of the commercial products or services referenced herein; nor should the mere reference herein to any drawing, specification, chemical process, or other data be regarded as a license or as a conveyance of any right or permission to the holder, reader, or any other person or corporation, to manufacture, reproduce, use, or sell any patented invention or copyrighted work that may in any way be related thereto. Registered names, trademarks, etc., used in this publication, even without specific indication thereof, are not to be considered unprotected by law.

PRINTED IN THE UNITED STATES OF AMERICA

ACS Symposium Series

M. Joan Comstock, *Series Editor*

1992 ACS Books Advisory Board

QP801
F45 A43
1991
CHEM

V. Dean Adams
Tennessee Technological
University

Mark Arnold
University of Iowa

David Baker
University of Tennessee

Alexis T. Bell
University of California—Berkeley

Arindam Bose
Pfizer Central Research

Robert Brady
Naval Research Laboratory

Dennis W. Hess
Lehigh University

Madeleine M. Joullie
University of Pennsylvania

Mary A. Kaiser
E. I. du Pont de Nemours and
Company

Gretchen S. Kohl
Dow-Corning Corporation

Bonnie Lawlor
Institute for Scientific Information

John L. Massingill
Dow Chemical Company

Robert McGorrin
Kraft General Foods

Julius J. Menn
Plant Sciences Institute,
U.S. Department of Agriculture

Vincent Pecoraro
University of Michigan

Marshall Phillips
Delmont Laboratories

A. Truman Schwartz
Macalaster College

John R. Shapley
University of Illinois
at Urbana–Champaign

Stephen A. Szabo
Conoco Inc.

Robert A. Weiss
University of Connecticut

Peter Willett
University of Sheffield (England)

Foreword

THE ACS SYMPOSIUM SERIES was founded in 1974 to provide a medium for publishing symposia quickly in book form. The format of the Series parallels that of the continuing ADVANCES IN CHEMISTRY SERIES except that, in order to save time, the papers are not typeset, but are reproduced as they are submitted by the authors in camera-ready form. Papers are reviewed under the supervision of the editors with the assistance of the Advisory Board and are selected to maintain the integrity of the symposia. Both reviews and reports of research are acceptable, because symposia may embrace both types of presentation. However, verbatim reproductions of previously published papers are not accepted.

Contents

THERMAL GENERATION

INDEXES

Preface

Fresh FRUITS AND VEGETABLES start with good flavor when harvested, but they increase or, more often, decrease in good flavor and sometimes develop undesirable flavors before reaching the consumer. Many processed foods develop desirable flavors from rather bland starting materials. As more knowledge of characteristic flavors is gained, understanding the formation and origin of flavor compounds becomes more important. The fundamental flavor chemistry and mechanisms by which flavor compounds are formed are important to scientists engaged in genetic engineering of plants and animals to improve flavor in the starting materials of food products. This knowledge is also needed in processing to produce food products of optimum flavor, and in storage and transportation to maintain flavor so that food products of the most acceptable quality may reach the consumers.

Books such as *The Maillard Reaction in Foods and Nutrition* (1983), *Biogeneration of Aromas* (1986), *Bioflavour '87* (1987), and *Thermal Generation of Aromas* (1989) are based on previous symposia covering similar topics as those in this book. However, when previous books were published, the topic of food flavoring compounds was of academic interest only. Today knowledge of flavor development mechanisms has some practical applications.

In this book, some of the most prominent flavor chemists in Australia, Europe, and the United States present the state of the art in flavor precursor chemistry.

ROY TERANISHI
Agricultural Research Service
U.S. Department of Agriculture
800 Buchanan Street
Albany, CA 94710

GARY R. TAKEOKA
Agricultural Research Service
U.S. Department of Agriculture
800 Buchanan Street
Albany, CA 94710

MATTHIAS GÜNTERT
Corporate Research
Haarmann and Reimer GmbH
Postfach 1253
D–3450 Holzminden, Germany

December 18, 1991

We dedicate this volume to peace.

–RT, MG, and GT–

Chapter 1

Thermal and Enzymatic Conversions of Precursors to Flavor Compounds

An Overview

Robert E. Erickson

Universal Flavors Corporation, 5600 W. Raymond, Indianapolis, IN 46241

This chapter gives an overview of the symposium and illustrates the commercial importance of thermal and enzymatic conversions of precursors to flavor compounds. In contrast to the flavor compounds found in fresh fruits and vegetables, others are formed during storage, drying, processing, and cooking of harvested foodstuffs. Animal products, which are usually consumed after thermal processing, have their own class of flavor compounds characteristic of meats. As more compounds related to flavor are identified, more studies are initiated to explain how these flavor compounds are developed and released. In this book, we have gathered papers by internationally known scientists who share the results of their investigations to add to the development of the complicated chemistry of food products.

The American food supply system is considered to be among the safest and best in the world. The flavor industry, as a part of that system, insures that consumers will have the aesthetic characteristics of palatability and appeal in their food products. The contribution of the flavor industry is not solely aesthetic, however, as enhanced palatability and appeal insure consumption with resulting nutritional benefits.

In the conversion of raw materials to finished food products, three major factors must be considered: economic feasibility, safety, and acceptability. After the first two elements are met, attention should be focussed on acceptability. Flavor, as well as color and texture, plays an important role in acceptability. Flavor has always been considered important in the food industry and has been and is dealt with empirically. Now it can be studied quantitatively on a scientific basis. As more characteristic flavor qualities are defined as mixtures of varying concentrations of specific compounds, it is of interest to learn the identity of the precursors of the compounds responsible for characteristic flavors,

0097–6156/92/0490–0001\$06.00/0
© 1992 American Chemical Society

and the conditions under which optimum amounts are obtained. The conversion of precursors to flavor compounds is not only important scientifically but also is of interest and economic value to processors and plant breeders. In this book some thermal and enzymatic conversions of precursors to flavor compounds are discussed.

The flavor industry began with the discovery that flavoring agents can be isolated by extraction and distillation from food materials and then used to impart a characteristic flavor to other food systems. For example, it was found that a lemon pudding can be made without lemons. The industry developed further as a branch of the fine chemical industry when it was discovered that isolated flavor materials can be chemically analyzed to identify individual components and then reassembled from the same components produced synthetically, usually from petroleum.

Today, the industry continues to make natural flavors by extraction and distillation, artificial flavors from synthetic aroma chemicals, and natural and artificial flavors from blends of the two. However, the industry also meets the needs of a consumer market in which there is increasing health awareness and preference for the natural. In order to meet these market needs, the industry must develop new science and technology which will allow the production of commercial quantities of safe, natural flavor ingredients. It is no longer sufficient to analyze and synthesize. Today we must understand and mimic nature and accurately reproduce the conditions of food processing.

The symposium on "Thermal and Enzymatic Conversions of Precursors to Flavor Compounds" addresses two key areas which have a great impact on the flavor industry. The first area is the use of biocatalysts, either as tools for the chemist to make specific natural ingredients or as reaction promoters in natural biochemical systems to generate flavor mixtures. The flavor generation in biological systems may be by biochemical reactions of appropriate precursors or by liberation of blocked or bound flavor ingredients. Examples of each are discussed in this book.

Enzymatic action has been known almost as long as the concept of catalysis. However, only in the last few decades has there been interest and application of enzymology in food science (1). Enzymatic action is occurring in foods from the time the raw material is grown and harvested until it is processed and shipped to the consumer's table. Improvements in analytical methods and identification of characteristic compounds have made it possible to follow the ripening of fruit (2, 3) and the rapid appearance and disappearance of some compounds (4, 5). Good examples of work in this area are given in "Biogeneration of Aromas" (6).

The second area covered in this book is the broad one of processed flavors and model systems in which certain carbohydrates, amino acids, lipids and vitamins are heated under conditions mimicking food processing to generate flavors by thermal reactions. Such flavors are considered to be natural by USA government regulations but may not be in other countries.

Since fire was discovered, raw materials have been exposed to high temperatures to form products which are considered highly desirable, such as breads, cakes, cookies, roasted meats and nuts, cocoa, coffee, etc. Chemical reactions involving thermal degradations and combinations of the fragments of carbohydrates, lipids, and amino acids have been and are currently being studied.

Investigations in forming flavors were made just as early as investigations in finding characteristic flavor compounds. One of the great pioneers in studying formation of flavor compounds was Louis-Camille Maillard. A brief history of Maillard and his accomplishments are given by S. Kawamura (7). The reaction of amino acids with sugars was named after Maillard and has been studied extensively, resulting in many books and symposia on the Maillard reaction [see Kawamura's review (7) for extensive references].

J. Hodge presented an outline of various reactions of the chemistry of browning reactions in model systems (8) and the origin of flavors in foods by nonenzymatic browning reactions (9). The information given in these reviews by Hodge is as pertinent and useful today as when originally written, and any serious researcher studying the Maillard reaction should occasionally review these excellent publications.

The complexity of products from heating processes is well known. S. Fors (10) has listed nearly 450 Maillard reaction products with some of their sensory properties. I. Flament (11) has reviewed the history of analytical research on coffee, cocoa, and tea, and has made comments on the structure, origin, and some sensory properties of the compounds present. Over 600 compounds have been identified in coffee volatiles, and over 400 in cocoa and tea volatiles (11). Not only are there many products formed from heat treatment, but also, there are many foods in which such compounds are important (12, 13).

Although much progress has been made, complete characterization of roasted foods and beverages is far from being accomplished. Some detailed investigations of products from heat treatment of carbohydrates, lipids, lignin, and amino acids and peptides are discussed in this book.

The term "precursor" is worthy of some discussion. The dictionary definition is simply "...thing that goes before, a predecessor". A chemical precursor is simply the substance going into a reaction which is modified according to the chemistry of the reaction. The modified precursor, the product, can undergo a second reaction for which it is now the precursor. This concept is important because nearly all natural flavor systems are dynamic with cascades of reactions in which products of one reaction become precursors for the next. The flavor of a fruit, for example, changes continually as it ripens and during post harvest storage. What then is the target fruit flavor for which the flavor industry must design in a man-made replacement? Only by detailed knowledge of the mechanisms of precursor generation and reaction can we intelligently design new and improved flavors.

It has been shown that glycosidically bound flavors can be precursors for the appropriate enzyme to generate the free flavor components. Examples are the flavor components of nectarine, grapes, and the limonoid components of citrus. Conversely, depleted pools of precursors can be refilled by external application of appropriate reactants. Both techniques present opportunities for commercial flavor preparation by enzymatically treating existing precursors or by supplying additional precursors to indigenous enzyme systems. The biosynthesis of flavor components in plants is demonstrated in the formation of lactones, various monoterpenes, and the unique group of C_{13} norisoprenoids.

Biocatalysts are superb tools for the flavor chemist because they can yield natural ingredients of appropriate chirality. The synthesis of d,l-carvone not only does not yield a natural ingredient but also mixes the distinct odor qualities of the two isomers. Chiral biocatalytic synthesis, on the other hand, yields a natural product of pure odor character. Examples of this type of synthesis are discussed in the papers on the use of carboxylesterase and lipases in organic solvents.

The area of processed flavors is extremely important to the flavor industry. The demand for low fat, yet tasty, meat and snack products requires the replacement of animal derived fat flavor systems with natural systems based on processed flavor development. Almost half the symposium was directed toward thermal generation of flavors, either in food systems or in model systems of mixtures of specific ingredients. Only by a more thorough understanding of the mechanisms of precursor and product modification in these systems can the industry hope to provide quality products.

A number of food systems that have been studied include the generation of the important flavor compound, furaneol. Specific categories of food components generate specific types of food flavors, such as carbonyl compounds from lipids and smoke flavors from lignin.

In five of the symposium papers the importance of model systems relating to the generation of processed flavors cannot go without comment. The systems involved in the thermal generation of flavors are extremely complex and present challenging, if not impossible, analytical problems. The design of a model system is a common approach in science to simplify analysis with full recognition that the relationship to reality is only as good as the design of the model. Many of the precursors which the industry has found to be useful components of processed flavor systems have been discovered through study of models such as those discussed in this symposium.

Examples of model systems which have been studied include ribose/ cysteine/thiamine mixtures, thiamine alone, peptides, cysteine/dextrose, and cysteine/sugars. Each of these model systems yielded valuable information relating to the formation of characteristic flavor compounds in food.

The importance of analytical techniques , mentioned above, is illustrated in the studies of "living" flavors and odors as well as in the Maillard chemistry of glucosones. In the history of the flavor industry, tremendous growth took place with the development of gas chromatography for the analysis of natural flavors from foods. Similar advances in knowledge of flavor precursors and their reactions will parallel the development of new analytical technology such as gas chromatography/mass spectrometry (GC/MS) and Fourier transform infrared spectrometry (FT/IR) coupled with capillary GC.

It should be noted that the generation of flavors in both biochemical and processing systems is governed by chemistry and the conditions of the process. The perception of the flavors by humans is followed by a judgment as to whether they are "good" or "bad". Good flavors are commercial products. Bad or off-flavors are food manufacturers' nightmares. Therefore, it is equally important to understand the conditions required to avoid bad flavors. The studies reported here help to develop an understanding of the mechanisms of flavor generation. This understanding can be applied to both "good" and "bad" flavors.

The advancement of our knowledge of the subjects covered in this symposium is significant to the flavor industry if we are to keep up with the changing demands of the marketplace. We have moved from the simple isolation of flavors to the analysis and reconstruction from natural and synthetic ingredients. It is now time to move on to a more complete understanding of the bio- and thermo- chemistries of our flavor generating systems. This is not only to improve quality but also to increase safety since, a priori, a system about which much is known is inherently safer than the unknown.

Literature Cited

1. Schwimmer, S. *Source Book of Food Enzymology*, The AVI Publishing Co., Inc., Westport, **1981**, 863 pp.
2. Dirinck, P., De Pooter, H., and Schamp, N. In *Flavor Chemistry: Trends and Developments,* Teranishi, R., Buttery, R. G., and Shahidi, F., Eds.; ACS Symposium Series 388; American Chemical Society, Washington, DC, **1989**, pp. 23-34.
3. Engel, K.-H., Ramming, D. W., Flath, R. A., and Teranishi, R. *J. Agric. Food Chem.* **1988**, *36 (5)*, 1003-1006.
4. Drawert, F., Kler, A., and Berger, R. G. *Lebensm.-Wiss. u.-Technol.* **1986**, *19,* 426.
5. Buttery, R. G., Teranishi, R., and Ling, L. C. *J. Agric. and Food Chem.,* **1987**, *35 (4)*, 540-544.
6. *Biogeneration of Aromas,* Parliment, T. H., and Croteau, R., Eds.; ACS Symposium Series 317; American Chemical Society, Washington, DC, **1986**, 397 pp.
7. Kawamura, S. In *The Maillard Reaction in Foods and Nutrition,* Waller, R. G., and Feather, M. S., Eds., ACS Symposium Series 215; American Chemical Society, Washington, DC, **1983**, pp. 3-18.
8. Hodge, J. E. *J. Agric. Food Chem.,* **1953**, *1 (15)*, 928-943.
9. Hodge, J. E. In *Symposium on Foods: The Chemistry and Physiology of Flavors*, Schultz, H. W., Day, E. A., and L.M. Libbey, Ed.; The AVI Publishing Company, Inc., Westport, **1967**, pp. 465-491.
10. Fors, S. In *The Maillard Reaction in Foods and Nutrition*, Waller, G. R., and Feather, M. S., Ed.: ACS Symposium Series 215: American Chemical Society, Washington, DC, **1983**, pp. 185-300.
11. Flament, I. *Food Reviews International*, **1989**, *5 (3)*, 317-414.

12.   *Thermal Generation of Aromas*; Parliment, T. H., McGorrin, R. J., and
      Ho, C.-H., Eds.; ACS Symposium Series 409, American Chemical
      Society, Washington, DC, **1989**, 548 pp.
13.   *The Maillard Reaction in Foods and Nutrition*, Waller, G. R., and
      Feather, M. S., Eds.; ACS Symposium Series 215, American Chemical
      Society, Washington, DC, **1983**, 585 pp.

RECEIVED December 18, 1991

ENZYMATIC REACTIONS

Chapter 2

Monoterpene Biosynthesis
Cyclization of Geranyl Pyrophosphate to (+)-Sabinene

R. B. Croteau

Institute of Biological Chemistry, Washington State University, Pullman, WA 99164–6340

The enzymatic conversion of geranyl pyrophosphate to the bicyclic olefin (+)-sabinene initiates a sequence of reactions responsible for the formation of the C3-oxygenated thujane family of monoterpenes in plants. A sabinene cyclase has been isolated from the leaves garden sage (*Salvia officinalis*), and a broad range of specifically-labeled substrates and substrate analogs has been employed to examine the reaction catalyzed by this enzyme. The results obtained define the overall stereochemistry of the coupled isomerization-cyclization of geranyl pyrophosphate (via 3*R*-linalyl pyrophosphate) to (+)-sabinene, demonstrate a 1,2-hydride shift in the construction of the thujane skeleton, and confirm the electrophilic nature of this enzymatic reaction type.

The thujane monoterpenes, including (+)-sabinene (**1**), (+)-3-thujone (**4**) and (-)-3-isothujone (**5**) (Scheme I), which bear the characteristic 1-isopropyl-4-methyl-bicyclo [3.1.0]hexane skeleton, represent one of the major classes of bicyclic monoterpenoids (*1*). Members of this family are widespread in the plant kingdom, and are found in the essential oils of many representatives of the Lamiaceae, Asteraceae, Rutaceae, Lauraceae, Verbenaceae, Piperaceae and Cupressaceae in which they contribute characteristic flavor and odor properties (*2*). (+)-3-Thujone and (-)-3-isothujone (components of the liqueur absinthe [*ex Artemisia* spp.] that was prepared in France before its prohibition in 1915) are among the more toxic of the naturally occurring monoterpenoids, and the antinociceptive activity (diminishing the response to pain) of (-)-3-isothujone is well-documented (*3*).

Elucidation of the biosynthesis of C3-oxygenated thujane monoterpenes (e.g., sabinone (**3**), thujone (**4**), isothujone (**5**)) has followed a circuitous path from an initial hypothesis involving the photooxidation of sabinene (**1**), to the suggested involvement of the monocyclic alcohols α-terpineol (see structure **10** for the corresponding cation) and terpinen-4-ol (**24**) as intermediates, to a proposal involving a series of 1,2-hydride shifts (*4*). Direct testing of [10-³H]sabinene, as well as [10-³H]α-thujene(the endocyclic Δ³,⁴-isomer of sabinene), [3-³H]terpinen-4-ol, and [3-³H]α-terpineol, in excised leaves of garden sage (*Salvia officinalis*;

0097–6156/92/0490–0008$06.00/0

© 1992 American Chemical Society

Lamiaceae), tansy (*Tanacetum vulgare*: Asteraceae), and wormword (*Artemisia absinthum*; Asteraceae) demonstrated that only sabinene (**1**), via *trans*-sabinol (**2**), was the precursor of thujone (or isothujone) in these plants, and this eliminated the other cyclic compounds as potential precursors (*5*) (for revised nomenclature of the thujane monoterpene, see Erman (*6*)). Furthermore, studies with cell-free extracts of these tissues established the conversion of geranyl pyrophosphate (**6**, Scheme II), the ubiquitous C_{10} intermediate of isoprenoid metabolism, to sabinene by an operationally soluble enzyme that seemingly carries out this complex cyclization reaction without free intermediates (*5*). Sabinene (the (+)-(1R, 5R)-enantiomer (**1**)) is transformed to (+)-(1S, 3R, 5R)-*trans*-sabinol (**2**) (for absolute configuration of the thujane monoterpenes, see Erman (*7*)) by a microsomal, NADPH/O_2-dependent cytochrome P-450 hydroxylase (*8*), and the alcohol is oxidized to (+)-sabinone (**3**) by a relatively specific, soluble dehydrogenase (*9*). Subsequent steps in the synthesis of C3-oxygenated thujane derivatives (acetylation of sabinol, stereoselective double-bond reduction of sabinone) have not yet been firmly established, but are likely to be carried out by well characterized enzyme types known to catalyze such transformations in the monoterpene series (*10-14*). The pathway to thujone and isothujone thus fits the general allylic oxidation-conjugate reduction scheme for the metabolism of monoterpene olefins (*15*). A similar sequence of reactions is probably involved in the transformation of α-thujene (thuj-3-ene) to the C2-oxygenated thujane derivative umbellulone (*15*).

Although the cyclization of geranyl pyrophosphate (**6**) to sabinene (**1**) is the key transformation in the construction of many thujane-type monoterpenes (*5*), and probably represents the rate-limiting step of such biosynthetic sequences (*16*), the enzyme has been rather neglected. Sabinene often constitutes a minor component of the essential oils (it being largely transformed to other products) and, since in early work the precursor role of this olefin was not appreciated, the cyclization was likely overlooked (*17*). Additionally, unless the appropriate carrier standards are added during the analysis of cyclization products, volatile olefins such as sabinene will be lost by evaporation during sample handling and chromatographic separation (*18*). Sabinene cyclase has been demonstrated in cell-free leaf extracts of sage and of various tissues of other plant species (*5*, *19*). The enzyme from sage and tansy leaf is operationally soluble, has a molecular weight of roughly 70,000, shows a pH optimum near neutrality, and requires a divalent metal ion for catalysis (Mg^{2+} preferred). Virtually nothing is known about the stereochemistry or mechanism of the cyclization leading to sabinene; however, a variety of approaches have been applied to understanding catalysis by other monoterpenoid cyclases (*15*). In this paper, I describe a range of these methods directed toward the cyclization to sabinene that constitutes a case study in monoterpene cyclization.

Isolation of Sabinene Cyclase

Sabinene cyclase was extracted from immature sage leaves by a standard protocol using 0.1 M sodium phosphate buffer, pH 6.5, containing reducing agents and the polymeric adsorbents polyvinylpolypyrrolidone and XAD-4 polystyrene resin to adsorb deleterious phenolic materials and oils, respectively (*18*). The soluble enzyme fraction was obtained by ultracentrifugation, and this supernatant was diluted and separated by anion-exchange chromatography on diethylaminoethyl-cellulose (*20*). The cyclase fraction (eluting at 450 mM KCl) was concentrated by ultrafiltration and subjected to gel permeation chromatography on Sephacryl S-200, a procedure that provided partial purification from competing activities such as phosphatases (sabinene cyclase elutes at a volume corresponding to a molecular

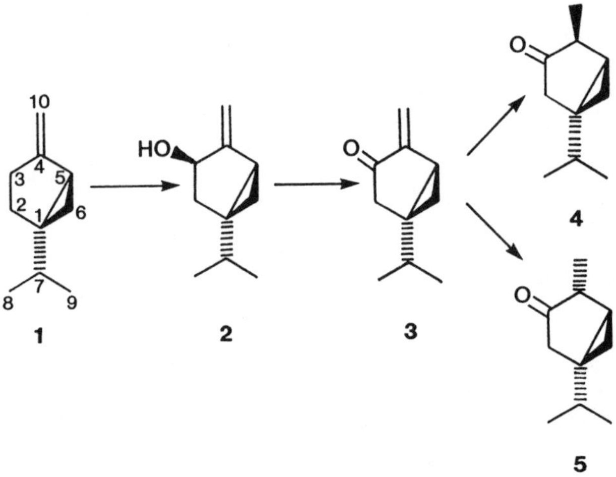

Scheme I.

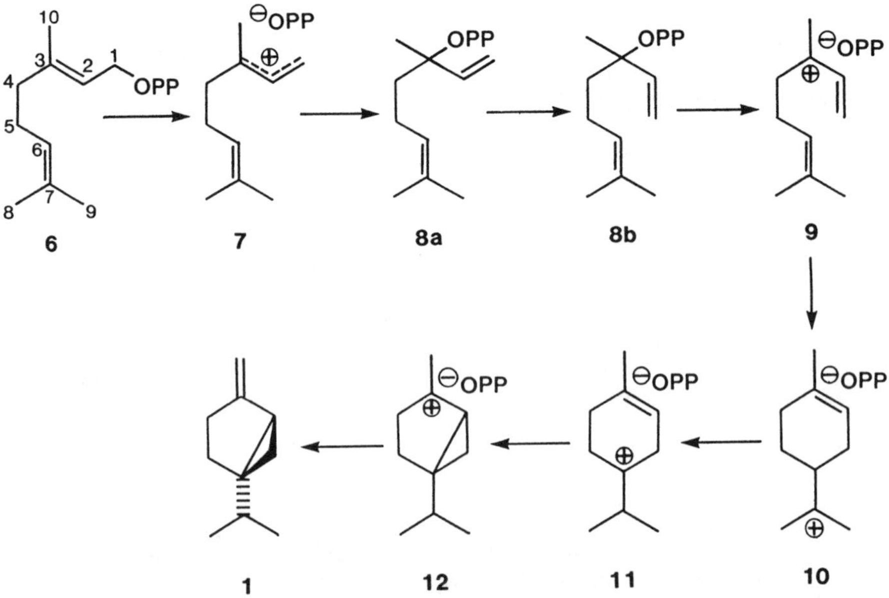

Scheme II. OPP indicates the pyrophosphate moiety.

weight of roughly 70,000) and allowed a change in buffer to assay conditions (20 mM Mes, pH 6.7, 5 mM KH_2PO_4, 15 mM $MgCl_2$, 0.5 mM dithiothreitol, and 10% (v/v) glycerol).

The assay for sabinene cyclase activity is based on the extraction of olefinic products into pentane, removal of oxygenated contaminants (terpenols) by column chromatography on silica gel, and, following the addition of internal standards, the purification of sabinene by argentation-TLC; linear assay conditions (by liquid scintillation counting of the purified product) were established using the standard substrate [1-^{3}H]geranyl pyrophosphate (*18*). General strategies and protocols for evaluating terpene cyclases are described in relevant reviews (*15, 18, 21*), and the preparation and use of the various labeled substrates and analogs are reported in the literature (these methods are referenced in the appropriate following sections).

Product Identity, Stereochemistry and Labeling Pattern

To establish the identity of the enzymatic product derived from [1-^{3}H]geranyl pyrophosphate in sage leaf extracts, the presumptive sabinene was prepared by large-scale incubation, the olefinic fraction was diluted with (±)-sabinene (prepared by synthesis (*22*) from α-terpinene), and the sabinene was isolated by argentation TLC. This material consisted of a single radioactive component coincident with the corresponding authentic sabinene standard on radio-GLC using two stationary phases of widely differing polarity (*23*). However, radio-chromatographic coincidence is insufficient to confirm product identity, and so the olefin was oxidized with ruthenium-tetraoxide (*24*) to sabina ketone (Scheme III, **13**) that was crystallized to constant specific activity as the semicarbazone.

Resolution of the biosynthetic product was achieved by conversion of sabinene (**1**) to sabinone (**3**) via selenium dioxide-PCC oxidation, followed by catalytic hydrogenation to 3-thujone (**4**), and radio-GLC separation of the diastereomeric ketals prepared from (*2R, 3R*)-2,3-butanediol (*25*). Radio-GLC demonstrated that only the (+)-(1*R*,5*R*)-antipode (**1**) was generated from [1-^{3}H]geranyl pyrophosphate by the sabinene cyclase from sage.

According to Ruzicka's original scheme for the biogenesis of the thujane monoterpenes (*26*), and to more recent mechanistic considerations (*27,28*), tritium from [1-^{3}H]geranyl pyrophosphate should label specifically the cyclopropyl methylene group (C6) of sabinene. Therefore, a degradative scheme was devised to generate a carbonyl function adjacent to this methylene group, thus rendering the corresponding α-protons susceptible to base-catalyzed exchange. Preparative scale incubation of [1-^{3}H,2-^{14}C]geranyl pyrophosphate (*29*) with the partially purified enzyme afforded (+)-sabinene of unchanged ^{3}H:^{14}C ratio that, following dilution with authentic carrier, was first transformed to terpinen-4-ol (**24**) (*30*) and then converted to 2-keto-1,4-cineole (Scheme III, **14**), via 2-iodo-1,4-cineole and 2-hydroxy-1,4-cineole, by an established sequence of reactions (*27*). The ^{3}H:^{14}C ratio of the derived ketone was essentially identical to that of both sabinene and the acyclic precursor (~ 8:1). However, exhaustive base treatment of the 2-keto-1,4-cineole removed approximately 95% of the tritium, confirming that the label from [1-^{3}H]geranyl pyrophosphate was located almost exclusively at C6 of sabinene. This result is fully consistent with the cyclization scheme involving participation of the double bond of the terpinen-4-yl intermediate (**11**) in the construction of the cyclopropane ring.

The sequence of steps leading from geranyl pyrophosphate (**6**), through the required isomerization to linalyl pyrophosphate (**8**), on to the monocyclic α-terpinyl cation (**10**) (Scheme II) is thought to be common to all monoterpene cyclizations

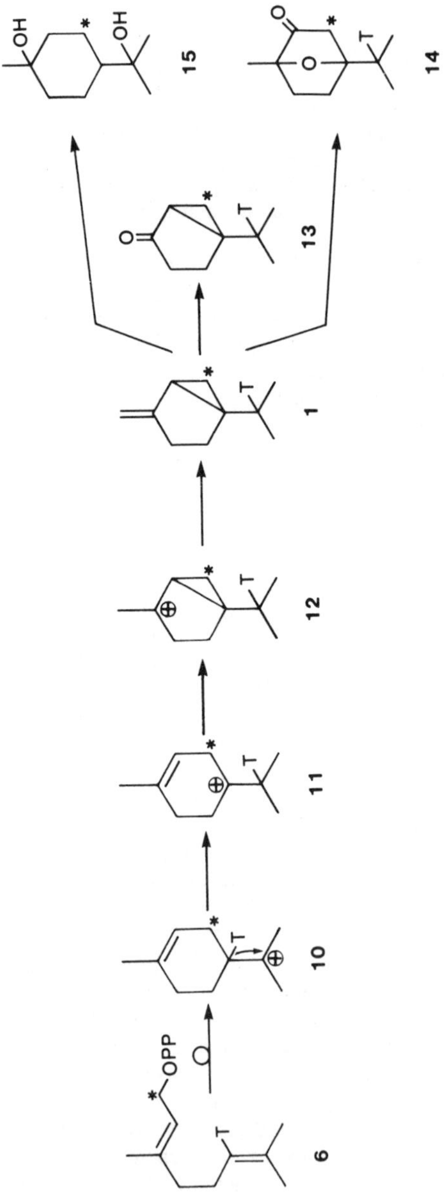

Scheme III. T represents ³H at C6, and the asterisk represents either ³H or ¹⁴C at C1, of geranyl pyrophosphate (**6**).

(15). For the formation of the thujane skeleton, a 1,2-hydride shift in the α-terpenyl cation (**10**) is required to provide the terpinen-4-yl cation (**11**) and thus permit cyclopropane ring closure via the cyclohexenyl double bond, consistent with the labeling pattern from [1-^{3}H]geranyl pyrophosphate. To examine the proposed 1,2-hydride shift, which is a key feature in the formation of the thujane family, the biosynthetic conversion of [6-^{3}H,1-^{14}C]geranyl pyrophosphate (**28**) to sabinene was exploited (Scheme III). Preparative scale conversion of the 6-^{3}H,1-^{14}C-labeled precursor (^{3}H:^{14}C=3.7) by the partially purified cyclase gave chromatographically pure (+)-sabinene of ^{3}H:^{14}C ratio nearly equal to that of the starting material (^{3}H:^{14}C=3.6), thus indicating that the ^{3}H at the C6-position must have undergone migration since C6 becomes substituted by only carbon in the transformation. The product was next diluted with authentic carrier, epoxidized and reduced to a mixture of *cis-* and *trans-*sabinene hydrate without change in ^{3}H:^{14}C ratio, and the mixture of epimeric sabinene hydrates was then oxidized with alkaline potassium permanganate to the corresponding diols (Scheme III, **15**) that were purified by TLC and each recrystallized to constant specific activity *(28)*. Examination of the radioisotope content of the products revealed essentially complete loss of ^{3}H from both the *cis-* and the *trans-*diol (**15**), thus establishing the migration of ^{3}H to C7 of the corresponding sabinene product, and confirming the proposed 1,2-hydride shift directly.

Ionization and Ion-Pairing

Both the isomerization and cyclization components of the enzymatic reaction sequence (Scheme II) are thought to occur at the same active site and to be initiated by the same event, ionization of the respective primary and tertiary allylic pyrophosphates to the corresponding ion pairs *(15)*. Evidence for the electrophilic nature of both steps has been obtained with 2-fluorogeranyl and (3*RS*)-2-fluorolinalyl pyrophosphate (Figure 1, **16** and **17**) as alternate substrates in which the electron-withdrawing fluorine substituent would be expected to retard ionization at the respective primary and tertiary centers *(31)*. A similar approach has been exploited by Poulter in deducing the electrophilic reaction mechanism of prenyltransferase, that involves a single ionization step *(32,33)*. The rate suppressions observed (roughly two orders of magnitude) using these fluorinated C_{10} analogs with the (+)-sabinene cyclase, by comparison with corresponding rates for solvolysis and nucleophilic displacement *(33)*, led to the conclusion that this enzyme functions by ionization of the relevant allylic pyrophosphate in both isomerization (geranyl pyrophosphate (**6**)) and cyclization (linalyl pyrophosphate (**8**)) steps and does not involve concerted displacements *(31)*.

The conclusions drawn with fluorinated analogs, which by competitive inhibition studies were shown to closely resemble the normal substrates in binding behavior ($K_i \sim K_m$), were bolstered by findings with other types of competitive inhibitors bearing the allylic pyrophosphate functionality, but differing markedly in the alkyl substituent *(34)*. Thus, the C_5 and C_{15} isoprenologues of geranyl pyrophosphate, dimethylallyl (**18**) and farnesyl pyrophosphate (**19**), not only inhibited cyclization to (+)-sabinene but also were themselves enzymatically solvolyzed at rates approaching those of the normal cyclization of geranyl pyrophosphate. Dimethylallyl pyrophosphate (**18**) was converted primarily to the corresponding alcohol at about 80% the rate of cyclization, whereas farnesyl pyrophosphate (**19**) was transformed to a mixture of farnesol (*trans, trans*) and *trans*-nerolidol (the C_{15} analog of linalool) at nearly 20% the rate of geranyl pyrophosphate cyclization. The non-cyclizable analog 6,7-dihydrogeranyl pyrophosphate (**20**) *(35)*, on the other hand, gave rise (at 50% the rate of

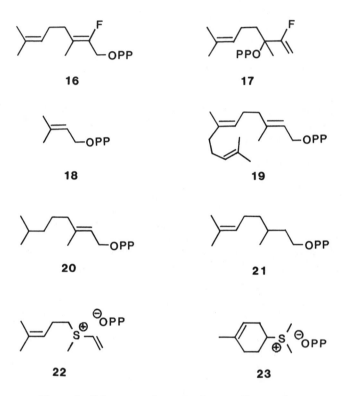

Figure 1. Substrate and reactive intermediate analogs.

cyclization) to a mixture of dihydrogeraniol, dihydronerol and dihydrolinalool (20% of total product) and to a mixture of the corresponding olefins (80% of total product). The relatively high proportion of acyclic olefins in this instance indicates ionization of the analog in a relatively hydrophobic environment in the enzyme active site, where deprotonation of the resulting cation(s) competes favorably with solvent capture. The effect likely results from the relatively tight binding of this analog ($K_m = 5$ µM) and the probability that the isopropyl group assumes the *endo* position normally occupied by the isopropylidene function of the natural substrate (see Scheme IV) and thus blocks premature backside water capture of the developing cation. The ability of monoterpene cyclases to catalyze the enzymatic solvolysis of allylic pyrophosphate substrate analogs (by contrast, 2,3-dihydrogeranyl pyrophosphate (**21**) is catalytically inactive) has thus provided an additional probe for the ionization steps of the reaction sequence (*34,35*).

The general proposal for monoterpene cyclization reactions involves the geranyl (**7**), linalyl (**9**) and α-terpinyl (**10**) cations as universal intermediates in the construction of six-membered ring derivatives (*15*) (Scheme II). The sulfonium ion analogs (**22 and 23**) of the two tertiary carbonium ion intermediates (**8 and 10**) previously have been shown to be potent reversible inhibitors of several monoterpene cyclases, and the inhibition is synergized by the presence of inorganic pyrophosphate, which is itself a substrate analog and competitive inhibitor (*28,36,37*). Such synergism indicates that these cyclases recognize and bind more tightly to the sulfonium ion•pyrophosphate anion paired species than to either ionic partner alone, and suggests that, in the course of the normal reaction, carbocationic intermediates remain paired with the pyrophosphate moiety as counterion (*36*).

Preliminary experiments with sabinene cyclase, employing kinetic methods described in detail elsewhere (*28,36*), confirmed that the sulfonium compounds, (*RS*)-methyl-(4-methylpent-3-en-1-yl)vinyl sulfonium perchlorate (**22**) and (*RS*)-dimethyl-(4-methylcyclohex-3-en-1-yl) sulfonium iodide (**23**), were inhibitors of the cyclization of geranyl pyrophosphate (K_i values of 0.2 µM and 0.7 µM, respectively, at a substrate concentration of 5 µM). Curiously, although the sulfonium analogs strongly inhibited the cyclization of geranyl pyrophosphate to sabinene, they promoted (to 2-3% of the normal rate of cyclization) the enzymatic solvolysis of the substrate to geraniol, a side reaction which is otherwise insignificant. The effect of inorganic pyrophosphate, as counterion, on inhibition by the sulfonium analogs was next examined. All combinations of inorganic pyrophosphate ($K_i \sim 120$ µM) and sulfonium analog afforded a typical synergistic response in that the mixtures yielded levels of inhibition much greater than the sum produced by the individual inhibitors as determined in control experiments. At the 5 µM geranyl pyrophosphate level, the observed K_i values for both sulfonium inhibitors were reduced fourfold by the presence of 50 µM inorganic pyrophosphate. The converse experiment was also carried out in which the apparent K_i value for inorganic pyrophosphate was shown to decrease by a factor of four in the presence of 0.1 µM of the linalyl analog (**22**) and by a factor of three in the presence of 0.35 µM of the α-terpinyl analog (**23**) (i.e., from 120 µM to 30 and 40 µM, respectively). These experiments clearly demonstrate the synergistic interaction between the sulfonium analogs and inorganic pyrophosphate, consistent with previous studies using these inhibitors with other monoterpene cyclases (*28,36,37*). These results provide strong suggestive evidence for the intermediacy of the predicted cationic species that the sulfonium analogs were intended to mimic, substantiate the electrophilic nature and course of the reaction, and implicate ion-paired intermediates in the isomerization-cyclization sequence (*15*).

Isomerization and Cyclization Steps

The mechanism of the multistep cyclization of geranyl pyrophosphate is considered to involve initial ionization of the pyrophosphate ester, in a manner analogous to the prenyltransferase reaction (38) in which a divalent metal ion is presumed to assist (39,40). This step is followed by stereospecific *syn*-isomerization to one or the other antipodal (3R)- or (3S)-linalyl intermediates (depending on the folding of the geranyl precursor) and rotation about the newly formed C2-C3 single bond. The suprafacial isomerization process to generate linalyl pyrophosphate (8) is a requirement imposed by the *trans*-geometry of the geranyl substrate (6) which is itself incapable of direct cyclization (Scheme IV). Subsequent ionization and cyclization of the cisoid, *anti endo*-conformer of the bound linalyl intermediate affords the corresponding monocyclic (4R)- or (4S)-α-terpinyl cation (10) (41) that, in the case of sabinene, undergoes a hydride shift to either enantiomeric conformer of the terpinen-4-yl cation (11). Internal addition via the endocyclic double bond would afford the corresponding (1R)- or (1S)-sabinyl cation (12) with net inversion at C4 of the original α-terpinyl system. The reaction is terminated by deprotonation at the C10 methyl of the sabinyl cation (12).

On the basis of these configurational and conformation considerations, and the absolute configuration of (+)-(1R,5R)-sabinene (7), the configuration of the cyclizing linalyl pyrophosphate intermediate is predicted to be 3R (assuming an *anti-endo* transition state precedented by chemical model reactions (42) and by other monoterpene cyclizations (15) (Scheme IV)). Since all cyclases thus far examined can utilize linalyl pyrophosphate as an alternate substrate (15,41), this prediction can be tested directly using the optically pure linalyl enantiomers (43). Kinetic constants were separately evaluated for (3R)- and (3S)-[1Z-^{3}H]linalyl pyrophosphate, and compared to those for geranyl pyrophosphate, using partially purified sabinene cyclase preparations. The rate of sabinene formation as a function of concentration of each linalyl antipode was determined, and in both cases a typical hyperbolic saturation curve was obtained. Analysis of the linear double-reciprocal plots provided an apparent K_m value of 4.3 μM for (3R)-linalyl pyrophosphate with V of 2.6 nmol/h•mg protein, and of 11.6 μM, with V of less than 0.2 nmol/h•mg protein for (3S)-linalyl pyrophosphate. Clearly, the 3R-enantiomer was a far more effective precursor of sabinene than was the 3S-enantiomer, as predicted, with a difference in the catalytic efficiency factor (V/K_m) of more than thirtyfold. A similar preference for the (3R)-linalyl pyrophosphate enantiomer was exhibited by the sabinene cyclase from tansy (*T. vulgare*). The utilization of the "unnatural" enantiomer of linalyl pyrophosphate at a slow rate is typical of monoterpene cyclases, and often leads to aberrant products (44).

Linalyl pyrophosphate has never been observed as a free intermediate in the cyclization of geranyl pyrophosphate to sabinene, or to any other cyclic monoterpene, and it is therefore presumed to remain bound to the enzyme active site in the time-frame of the coupled reaction sequence (15,45). For this reason, the isomerization step of these coupled reactions cannot be viewed directly. Information about the relative rates of the two steps can be gained, however, by comparing kinetic constants for (3R)-linalyl pyrophosphate to those for geranyl pyrophosphate obtained under the same conditions (K_m = 2.2 μM, V = 0.54 nmol/h•mg protein). Since the enzyme encounters linalyl pyrophosphate only as a consequence of binding and isomerizing geranyl pyrophosphate, the calculated response to the solution concentration of (3R)-linalyl pyrophosphate is irrelevant, and V_{rel} is the more suitable parameter for comparative purposes. In this case, the cyclization of (3R)-linalyl pyrophosphate is about five times more efficient than is the coupled isomerization-cyclization of geranyl pyrophosphate, indicating that the isomerization of geranyl pyrophosphate is the slow step of the reaction sequence.

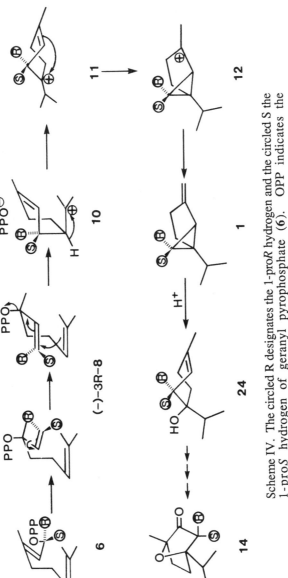

Scheme IV. The circled R designates the 1-proR hydrogen and the circled S the 1-proS hydrogen of geranyl pyrophosphate (6). OPP indicates the pyrophosphate moiety.

That linalyl pyrophosphate is cyclized at the same active site as geranyl pyrophosphate was confirmed by the mixed substrate method of Dixon and Webb (46). Thus, when $(3R)$-[1Z-³H]linalyl pyrophosphate was mixed in varying proportions with [1-³H]geranyl pyrophosphate, the velocities were observed to fall short of the expected sum for both substrates. These results indicate mutually competitive utilization of both precursors at a common active site. That the isomerization and cyclization occur at the same site is entirely consistent with the observed tight coupling of these reaction steps.

With the intermediacy of $(3R)$-linalyl pyrophosphate in the transformation firmly established, it became possible to examine in greater detail the proposed stereochemical model for the cyclization to (+)-$(1R,5R)$-sabinene and, in particular, to test the prediction that configuration at C1 of the geranyl substrate will be retained in the coupled reaction sequence. Thus, syn-isomerization of geranyl pyrophosphate to $(3R)$-linalyl pyrophosphate, followed by transoid to cisoid rotation about the newly generated C2-C3 single bond, brings the face of C1, from which the pyrophosphate moiety has departed, into juxtaposition with the isopropylidene function from which C1-C6 ring closure to the $(4R)$-α-terpinyl skeleton occurs (Scheme IV). To examine this crucial stereochemical prediction, (1S)- and (1RS)-[1-³H,2-¹⁴C]geranyl pyrophosphate (³H:¹⁴C ratio of both = 16) (29) were enzymatically converted to (+)-$(1R,5R)$-sabinene by the partially purified cyclase preparation (without loss of tritium). The enzymatically derived sabinene samples (³H:¹⁴C ~ 16) were diluted with authentic carrier and each then stereoselectively converted to 2-keto-1,4-cineole (14) as before (Scheme IV). The ³H:¹⁴C ratios of the derived ketones were unchanged from the starting materials as expected, allowing advantage to be taken of the selectivity of base-catalyzed exo-α-hydrogen exchange of this oxabicyclo[2.2.1]heptanone system in locating the tritium (the exo:$endo$ exchange ratio was determined to be 11.2:1 by ¹H-NMR monitoring of exchange of the bis-deuteroketone as previously described (28)).

Individual exchange runs with each sample of 2-keto-1,4-cineole derived from (1S)- and (1RS)-labeled geranyl pyrophosphate were carried out under controlled conditions (28) and tritium loss was determined relative to the ¹⁴C-labeled internal standard at 5 min intervals. Corrections were made to adjust for the 22% racemization which occurred during the preparation of 2-keto-1,4-cineole (28) and the exchange curves for the products derived from both geranyl precursors were plotted. The loss of tritium from 2-keto-1,4-cineole derived from (1S)-[1-³H,¹⁴C]geranyl pyrophosphate was much slower than that from the product derived from the corresponding racemically labeled precursor, indicating that the 1-pro$S$-hydrogen of the acyclic precursor gives rise predominantly to the $endo$-α-hydrogen of the derived keto-cineole (Scheme IV). Comparison of exchange rates for the products derived from (1RS)- and (1S)-labeled precursors gave a ratio of 5.3:1, which agrees well with prediction based upon the relative exo versus $endo$ exchange rates determined by ¹H-NMR monitoring of the bis-deuterated ketone exchanged under the same conditions (i.e., an absolute exo:$endo$ exchange ratio of 11.2:1 determined by NMR methods predicts an observed $racemate$ (1-RS-³H-derived):$endo$ (1-S-³H-derived) exchange ratio of 5.6:1, since the rate of ³H loss from the racemate-derived product would be halved).

From this data, it is clear that the configuration at C1 of geranyl pyrophosphate is retained in the enzymatic conversion to (+)-sabinene, an observation entirely consistent with the reaction scheme involving syn-isomerization and $anti$-cyclization. When taken together with the demonstrated configurational preference for $(3R)$-linalyl pyrophosphate, this finding supports the proposed helical conformation of the reacting geranyl precursor and the $anti,endo$

conformation of the cyclizing tertiary intermediate. The summation of these studies thus defines all of the major stereochemical elements in the isomerization and cyclization of geranyl pyrophosphate to (+)-sabinene. This evidence, along with the demonstration of the electrophilic nature of the reaction and the delineation of the 1,2-hydride shift, establish the key mechanistic features in the construction of (+)-sabinene and further support the general model for monoterpene cyclization (*15,47*).

Acknowledgments

This investigation was supported in part by National Institutes of Health Grant GM-31354 and National Science Foundation Grant DCB 91-04983, and by Project 0268 from the Washington State University Agricultural Research Center, Pullman, WA 99164. The expert assistance from N. M. Felton, C. J. Wheeler, D. M. Satterwhite, T. W. Hallahan and F. Karp is gratefully acknowledged.

Literature Cited

1. Whittaker, D.; Banthorpe, D. V. *Chem. Rev.* **1972**, *72*, 305-313.
2. Srinivas, S. R. *Atlas of Essential Oils*; Anadams: New York, NY, 1986, pp 1-826.
3. Rice, K. C.; Wilson, R. S. *J. Med. Chem.* **1976**, *19*, 1054-1057.
4. Banthorpe, D. V.; Charlwood, B. V.; Francis, M. J. O. *Chem. Rev.* **1972**, *72*, 115-155.
5. Karp, F.; Croteau, R. *Arch. Biochem. Biophys.* **1982**, *216*, 616-624.
6. Erman, W. F. In *Chemistry of the Monoterpenes: An Encyclopedic Handbook*; Gassman, P. G., Ed.; Studies in Organic Chemistry; Marcel Dekker: New York, NY, 1985, Vol. 11, Part B; pp 815-860.
7. Erman, W. F. In *Chemistry of the Monoterpenes: An Encyclopedic Handbook*; Gassman, P. G., Ed.; Studies in Organic Chemistry; Marcel Dekker: New York, NY, 1985, Vol. 11, Part A; pp 184-197.
8. Karp, F.; Harris, J. L.; Croteau, R. *Arch. Biochem. Biophys.* **1987**, *256*, 179-193.
9. Dehal, S. S.; Croteau, R. *Arch. Biochem. Biophys.* **1987**, *258*, 287-291.
10. Croteau, R.; Hooper, C. L. *Plant Physiol.* **1978**, *61*, 737-742.
11. Kjonaas, R.; Martinkus-Taylor, C.; Croteau, R. *Plant Physiol.* **1982**, *69*, 1013-1017.
12. Kjonaas, R. B.; Venkatachalam, K. V.; Croteau, R. *Arch. Biochem. Biophys.* **1985**, *238*, 49-60.
13. Croteau, R.; Venkatachalam, K. V. *Arch. Biochem. Biophys.* **1986**, *249*, 306-315.
14. Croteau, R.; Wagschal, K. C.; Karp, F.; Satterwhite, D. M.; Hyatt, D. C.; Skotland, C. B. *Plant Physiol.* **1991**, *96*, 744-752.
15. Croteau, R. *Chem. Rev.* **1987**, *87*, 929-954.
16. Gershenzon, J.; Croteau, R. In *Biochemistry of the Mevalonic Acid Pathway to Terpenoids*; Towers, G. H. N.; Stafford, H. A., Eds.; Recent Advances in Phytochemistry; Plenum Press: New York, NY, 1990, Vol. 24; pp 99-160.
17. Gambliel, H.; Croteau, R. *J. Biol. Chem.* **1982**, *257*, 2336-2342.
18. Croteau, R.; Cane, D. E. In *Steroids and Isoprenoids*; Law, J. H.; Rilling, H. C., Eds.; Methods in Enzymology; Academic Press: New York, NY, 1985, Vol. 110, pp 383-405.
19. Chayet, L.; Rojas, C.; Cardemil, E.; Jabalquinto, A. M.; Vicuna, R.; Cori, O. *Arch. Biochem. Biophys.* **1977**, *180*, 318-327.
20. Lanznaster, N.; Croteau, R. *Protein Express. Purif.* **1991**, *2*, 69-74.

21.   Alonso, W. R.; Croteau, R. In *Enzymes of Secondary Metabolism*; Lea, P. J., Ed.; Methods in Plant Biochemistry; Academic Press: London, UK, 1991, Vol. 9; in press.
22.   Fanta, W. I.; Erman, W. F. *J. Org. Chem.* **1968**, *33*, 1656-1658.
23.   Satterwhite, D. M.; Croteau, R. B. *J. Chromatogr.* **1988**, *452*, 61-73.
24.   Carlsen, H. J.; Katsuki, T.; Martin, V. S.; Sharpless, K. B. *J. Org. Chem.* **1981**, *46*, 3936-3938.
25.   Satterwhite, D. M.; Croteau, R. B. *J. Chromatogr.* **1987**, *407*, 243-252.
26.   Ruzicka, L.; Eschenmoser, A.; Heusser, H. *Experientia* **1953**, *9*, 362-367.
27.   Hallahan, T. W.; Croteau, R. *Arch. Biochem. Biophys.* **1988**, *264*, 618-631.
28.   Hallahan, T. W.; Croteau, R. *Arch. Biochem. Biophys.* **1989**, *269*, 313-326.
29.   Croteau, R.; Felton, N. M.; Wheeler, C. J. *J. Biol. Chem.* **1985**, *260*, 5956-5962.
30.   Cooper, M. A.; Holden, C. M.; Loftus, P.; Whittaker, D. *J. Chem. Soc. Perkin I* **1973**, 665-667.
31.   Croteau, R. *Arch. Biochem. Biophys.* **1986**, *251*, 777-782.
32.   Poulter, C. D.; Argyle, J. C.; Mash, E. A. *J. Biol. Chem.* **1978**, *253*, 7227-7233.
33.   Poulter, C. D.; Wiggins, P. L.; Le, A. T. *J. Am. Chem. Sec.* **1981**, *103*, 3926-3927.
34.   Wheeler, C. J.; Croteau, R. *J. Biol. Chem.* **1987**, *262*, 8213-8219.
35.   Wheeler, C. J.; Croteau, R. *Arch. Biochem. Biochem.* **1986**, *246*, 733-742.
36.   Croteau, R.; Wheeler, C. J.; Aksela, R.; Oehlschlager, A. C. *J. Biol. Chem.* **1986**, *261*, 7257-7263.
37.   Croteau, R.; Miyazaki, J. H.; Wheeler, C. J. *Arch. Biochem. Biophys.* **1989**, *269*, 507-516.
38.   Poulter, C. D.; Rilling, H. C. In *Biosynthesis of Isoprenoid Compounds*; Porter, J. W.; Spurgeon, S. L., Eds.; Wiley: New York, NY, 1981, Vol. 1; pp 161-224.
39.   Brems, D. N.; Rilling, H. C. *J. Am. Chem. Soc.* **1977**, *99*, 8351-8352.
40.   Chayet, L.; Rojas, C. M.; Cori, O.; Bunton, C. A.; McKenzie, D. C. *Bioorg. Chem.* **1984**, *12*, 329-338.
41.   Croteau, R. In *The Metabolism, Structure, and Function of Plant Lipids*; Stumpf, P. K.; Mudd, J. B.; Nes, W. D., Eds.; Plenum Press: New York, NY, 1987, pp 11-18.
42.   Gotfredsen, S.; Obrecht, J. P.; Arigoni, D. *Chimia*, **1977**, *31*, 62-63.
43.   Satterwhite, D. M.; Wheeler, C. J.; Croteau, R. *J. Biol. Chem.* **1985**, *260*, 13901-13908.
44.   Croteau, R.; Satterwhite, D. M. *J. Biol. Chem.* **1989**, *264*, 15309-15315.
45.   Wheeler, C. J.; Croteau, R. *Proc. Natl. Acad. Sci. USA* **1987**, *84*, 4856-4859.
46.   *Enzymes*, 3rd ed.; Dixon, M.; Webb, E. C., Eds.; Academic Press, New York, NY, 1979, pp 72-75.
47.   Cane, D. E. *Acc. Chem. Res.* **1985**, *18*, 220-226.

RECEIVED December 18, 1991

Chapter 3

Lipases: Useful Biocatalysts for Enantioselective Reactions of Chiral Flavor Compounds

Karl-Heinz Engel

Institut für Biotechnologie, Technische Universität Berlin, Fachgebiet Chemisch-Technische Analyse, Seestrass 13, D–1000 Berlin 65, Germany

A commercially available lipase preparation from *Candida cylindracea* (CCL) has been employed for kinetic resolutions of the enantiomers of chiral acids and their esters.   CCL-catalyzed esterification of 2-methylalkanoic acids and transesterification of the corresponding esters via acidolysis proceed with a preference for the (S)-configurated substrates.   Structural influences on velocity and enantioselectivity of these reactions are demonstrated. Limitations of the acidolysis due to the reversibility of the reaction are discussed.   Lipase-catalyzed reactions of hydroxyacid esters with fatty acids in organic solvent are characterized by a competition of esterification and transesterification. The rates of these pathways and their stereochemical course depend on the position of the hydroxy group and determine the final distribution of products.

Lipases (triacylglycerol acylhydrolases, E.C. 3.1.1.3) play an important role among the various enzymes and microorganisms employed as biocatalysts in organic synthesis (*1*).   The extension of their use beyond the traditional field of fat modification (*2*) has been mainly due to their ability (i) to accept a diverse spectrum of "unnatural" substrates different from triglycerides, (ii) to catalyze reactions not only in aqueous but in organic media (*3,4*), and (iii) to differentiate between enantiomers if chiral substrates are employed (*5,6*). Based on these outstanding features lipases are also being increasingly used in the area of flavor and aroma chemistry to synthesize "natural" flavoring materials such as esters (*7,8*) and to obtain chiral flavor compounds in optically pure form via kinetic resolution of racemic precursors (*9,10*).

We have been investigating lipases in the course of our continuous studies on chiral volatiles in plants and microorganisms (*11,12*).   Commercially available enzyme preparations were screened as potential sources for biotechnological production of optically active volatiles; in addition, they served as model systems for investigations of biogenetic pathways leading to chiral flavor compounds.   Results

0097–6156/92/0490–0021$06.00/0
© 1992 American Chemical Society

obtained for lipase-catalyzed reactions of chiral acids and esters, respectively, bearing the asymmetric center in their acyl moiety will be presented in this chapter.

Lipase-catalyzed reactions of 2-methyl branched acids

2-Methyl branched acids and their esters are important aroma contributing constituents of various natural systems, especially fruits (13,14).   An almost exclusive presence of the (S)-enantiomers of 2-methylbutanoic acid and ethyl 2-methylbutanoate in pineapple and strawberry has been demonstrated (15, 16). Determination of odor thresholds and calculation of odor units revealed the outstanding role of ethyl (S)-2-methylbutanoate in the aroma of pineapples (17). A further need for optically pure 2-methyl branched acids and esters, respectively, is due to their use as synthons for other biologically active compounds, such as pheromones (18).

Approaches. Enantioselective lipase-catalyzed esterification is a strategy successfully applied to the kinetic resolution of the enantiomers of chiral alcohols (19,20). The use of this approach for chiral 2-substituted acids had been limited to substrates with electronegative substituents in the 2- position, such as 2-chloro- and 2-bromoalkanoic acids (21) or 2-(4-chlorophenoxy)propanoic acid (22). Lipase-catalyzed esterification of 2-methyl branched acids has been reported (23,24); however, data on a kinetic resolution of enantiomers had been lacking.

Lipase-catalyzed transesterification via acidolysis is a useful methodology to modify triglycerides (2). Due to a directed acyl exchange valuable lipids with desired properties which may be unobtainable by chemical reactions can be "tailored". The conversion of palm oil fractions into cocoa butter equivalents is a well-studied example (25). Lipases also accept other (unnatural) esters as substrates for acidolyses (26), but a potential enantioselectivity had not been investigated for this type of reaction.

Recently, our own studies (27,28) and those of two other research groups (29,30) demonstrated that lipase-catalyzed hydrolysis, esterification and transester-ification via acidolysis are useful strategies for kinetic resolution of the enantiomers of chiral 2-methyl branched acids and their esters, respectively.   The reaction schemes for esterification and acidolysis carried out in heptane as solvent and employing the lipase from Candida cylindracea (now called Candida rugosa) as biocatalyst are shown in Figure 1. In both processes the (S)-enantiomer is the preferred substrate. The kinetic and thermodynamic parameters determining the applicability of these reactions will be discussed.

Analytical Methodology . Due to the use of an organic solvent the time course of the reactions presented in Figure 1 could be monitored directly by means of periodic gas chromatographic analysis. Liquid-solid chromatography on silica gel was used to separate the product from the remaining substrate. Their optical purities were determined via capillary gas chromatographic separations of diastereomeric (R)-1-phenylethylamides and (S)-2-octyl esters, respectively. Enantiomeric excesses of both the product and the remaining substrate as determined at various conversion rates were used to calculate the enantioselectivity (E) of the reaction (31). Some of the data obtained for the CCL-catalyzed esterification of 2-methylpentanoic acid is presented in Figure 2.

Structural effects. Both procedures presented in Figure 1 are characterized by a consistent preference for the (S)-configured substrates irrespective of structural modifications (*27,28,30*). This principle has been observed for other lipase preparations and may be general for reactions of 2-methyl branched substrates; the only exception reported is the inverted stereoselection of the hydrolysis of 2-methyloctanoates by the lipase from *Rhizomucor miehei* caused by a change of the alcohol moiety from an aliphatic to an aryl alcohol (*29*).

On the other hand, structural variations have a strong impact on reaction rates and degree of enantiodiscrimination; these influences have been investigated in detail (*27,28,30*). A selection of data demonstrating major effects is presented in Table I. Substrate combinations (2-methyl branched acid/alcohol) employed for CCL-catalyzed esterifications are compared to those (corresponding 2-methylbranched ester/acid) used for the acidolysis reaction.

Table I: Comparison of reaction rates and enantioselectivities of esterifications and transesterifications catalyzed by lipase from *Candida cylindracea*

Substrates		Relative rate (%)	Enantioselec- tivity E[a]
2-methylbutanoic acid	ethanol	100[b]	3
ethyl 2-methylbutanoate	octanoic acid	46	6
2-methylbutanoic acid	octanol	76	4
octyl 2-methylbutanoate	octanoic acid	37	7
2-methylhexanoic acid	ethanol	13	1.4
ethyl 2-methylhexanoate	octanoic acid	11	13
ethyl 2-methylhexanoate	oleic acid	20	12
2-methylhexanoic acid	octanol	22	20
octyl 2-methylhexanoate	octanoic acid	7	52
2-methylhexanoic acid	octadecanol	28	30
octadecyl 2-methyl- hexanoate	octanoic acid	3	126

[a]cf. ref. (*31*). [b]Reaction rate v = 0.1 μMol·h^{-1}·mg^{-1}

The influences can be summarized as follows:

(a) An increase of the chain length of the alcohol substrate or the alcohol moiety of the ester substrate improves the enantioselectivities of esterification and acidolyis, respectively. For acidolyses this enhancement of enantiodiscrimination is accompanied by decreasing reaction rates. The poor stereoselection observed for reactions of 2-methylbutanoic acid and its esters cannot be influenced by structural variations of the alcohol.

(b) Modifications of the acid substrates have little impact on the enantio-selective course of the acidolysis; the choice of the acid is only important in terms

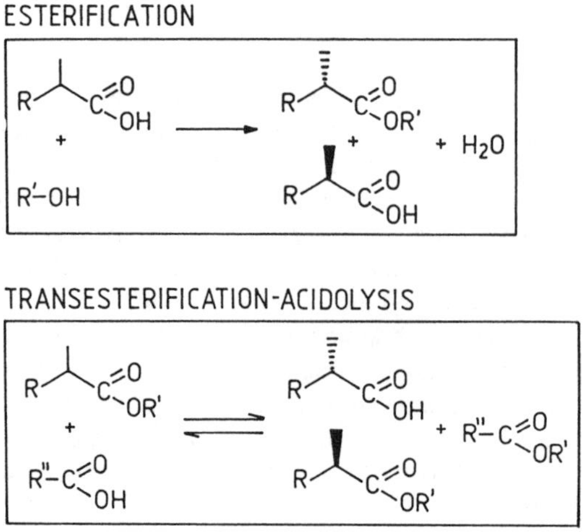

ESTERIFICATION

TRANSESTERIFICATION-ACIDOLYSIS

Figure 1: Kinetic resolution of the enantiomers of 2-methyl branched substrates via CCL-catalyzed reactions in heptane

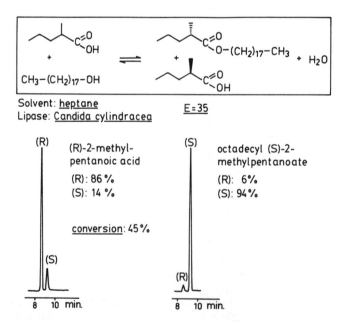

Solvent: heptane
Lipase: Candida cylindracea

E = 35

(R) (R)-2-methyl-
pentanoic acid
(R): 86 %
(S): 14 %

conversion: 45 %

(S)

8   10 min.

(S) octadecyl (S)-2-
methylpentanoate
(R): 6 %
(S): 94 %

(R)

8   10 min.

Figure 2: Optical purities determined during the course of the CCL-catalyzed esterification of 2-methylpentanoic acid. (Cf. ref. (27,28) for GC-separation of diasteromeric (S)-2-octyl esters)

of reaction rate. On the other hand, chain elongation of the 2-methyl branched acid moieties in the ester substrates employed for acidolysis and the acid substrates used for esterification significantly increases the enantiodiscrimination by the enzyme.

(c) CCL-catalyzed esterifications proceed faster than the corresponding acidolyses, but the enantiodiscrimination achieved by means of acidolysis of an ester substrate is significantly higher than that observed for the lipase-catalyzed synthesis of such an ester.

Thermodynamic parameters. CCL-catalyzed acidolysis of 2-methyl branched subtrates is superior to the esterification approach in terms of enantiodiscrimination. However, the applicability of this strategy is limited by the reversible nature of the underlying reaction. CCL-mediated esterifications of 2-methyl branched acids in organic solvent proceed nearly irreversibly; equilibrium conversions up to 98 % were determined (30). On the other hand, yields obtained in the course of lipase-catalyzed acidolyses remained below the theoretical values. For example, the reaction between octyl 2-methylhexanoate and oleic acid reached an equilibrium state after about 40 % conversion.

Figure 3 displays the absolute amounts of the enantiomers formed during the course of this reaction. The time course is typical for enzyme-catalyzed kinetic resolutions based on reversible reactions: The first stage is characterized by a preferred reaction of the (S)-ester; however, the acidolysis of this enantiomer approaches an equilibrium. The equilibrium constant can be calculated from the concentrations of remaining substrate and product of this faster reacting species at the equilibrium stage. Finally, the ongoing reaction of the (R)-ester causes decreasing optical purities of product and remaining substrate.

Both enantioselectivity (E) of the enzyme and equilibrium constant (K) of the reaction determine the enantiomeric excesses obtainable by means of lipase-catalyzed kinetic resolutions via acidolysis. These effects are demonstrated by the reaction profiles presented in Figure 4. The experimentally determined optical purities of product and remaining substrate are in good agreement with theoretical curves simulating the resolution process (32,33). The initially obtainable enantiomeric excess of the product and the time when the reaction has to be stopped to assure maximum optical purity of the remaining substrate are strongly influenced by the structural variations. Due to the low enantioselectivity and the unfavorable equilibrium the acido-lysis of ethyl 2-methylbutanoate would be of no significant value for preparative applications. On the other hand, the reaction of octadecyl 2-methyl-hexanoate demonstrates the potential of this strategy for kinetic resolution of the enantiomers of 2-methyl branched substrates.

Biocatalyst. The use of easily accessible and relatively inexpensive enzymes is one of the prerequisites for the application of the above described procedures on a large scale. Commercially available lipases fulfill these requirements; however, the enantioselectivity of these preparations may strongly depend on their purity. We noticed significant variations for different lots of the lipase from *Candida cylindracea* (Sigma, L 1754). Results obtained for the esterification of 2-methyl branched acids with octanol are presented in Table II.

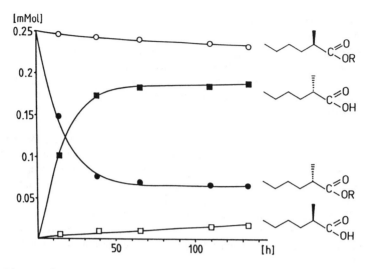

Figure 3: Amounts of the enantiomers formed during the CCL-catalyzed reaction between octyl 2-methylhexanoate and oleic acid

● = octyl (S)-2-methylhexanoate
○ = octyl (R)-2-methylhexanoate
■ = (S)-2-methylhexanoic acid
□ = (R)-2-methylhexanoic acid

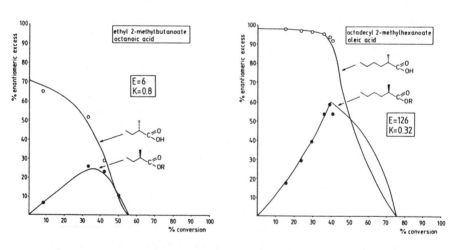

Figure 4: Relationship between conversion rate and enantiomeric excesses of product (e.e.P) and remaining substrate (e.e.S) for CCL-catalyzed acidolyses. For simulation of the graphs see (28,32).
○ = experimentally determined values for e.e.P
● = experimentally determined values for e.e.S

Table II: Enantioselectivities of different lots of lipase from *Candida cylindracea* in the course of the esterification of 2-methyl branched acids

	Enantioselectivity (E) lipase lot		
Acid[a]	(a)	(b)	(c)
2-methylpentanoic	20	14	8
2-methylhexanoic	70	33	17

[a]alcohol substrate: octanol

For the hydrolysis of structurally related arylpropanoic and phenoxy propanoic acid esters the strong impact of a purification step involving a treatment of the crude lipase from *Candida cylindracea* with deoxycholate and organic solvents on its enantioselectivity has been demonstrated (*34*).

Further studies on the protein structure, the enzyme-substrate complex, and the conformational environment required at the active site of the enzyme may be necessary to assure maximum and reproducible enantioselectivity of the biocatalyst.

Lipase-catalyzed reactions of hydroxyacid esters

Hydroxyacid esters and their acetoxy derivatives are characteristic volatile constituents of various tropical fruits. For this class of compounds an important principle later confirmed for other constituents has been demonstrated for the first time: chiral flavor and aroma compounds do not always occur in natural systems in optically pure form, but well-defined mixtures of enantiomers can be present (*35,36,37*). For hydroxyacid esters such enantiomeric ratios may be rationalized by a competition of enzymes and/or pathways involved in their biosynthesis (*12*). In order to elucidate some of the principles underlying the formation of optically active acyloxyacid derivatives we decided to study lipase-catalyzed reactions between racemic hydroxyacid esters and fatty acids in organic solvent as model systems (*38*).

Competing Pathways. Investigations of CCL-catalyzed reactions of ethyl 2-, 3-, and 5-hydroxyhexanoates revealed that these bifunctional compounds serve as substrates for two simultaneous pathways: (a) the formation of an acyloxyacid ester via esterification of the hydroxy group, and (b) the liberation of a free hydroxyacid via acidolysis of the ester employed (Figure 5). The final distribution of products and their optical purities are determined by a competition of these pathways. Both proceed enantioselectively; however, the rate and stereochemical course are strongly dependent on the position of the hydroxy group in the ester substrate.

In Figure 6 data is presented which was obtained from the CCL-catalyzed reactions between hydroxyacid esters and octanoic acid. Starting from racemic ethyl 2-hydroxyhexanoate the liberation of (S)-2-hydroxyhexanoic via acidolysis is the major reaction; the (R)-2-acyloxyacid ester is produced in minor quantities. Increasing the distance between the hydroxy and the carboxylic group in the ester substrate favors the esterification; only a minor part of ethyl 3-hydroxyhexanoate undergoes transesterification. In addition the shift of the asymmetric center causes an inversion of the stereochemical course of these competing pathways.

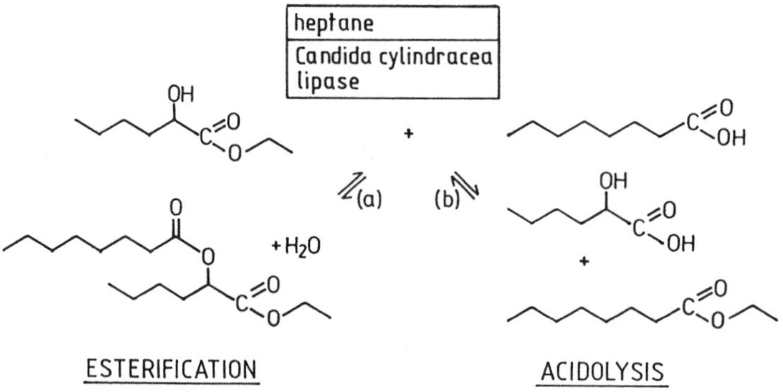

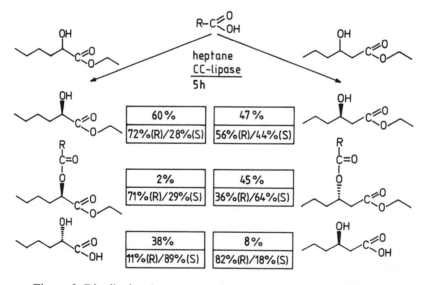

Figure 5:CCL-catalyzed reactions between ethyl 2-hydroxyhexanoate and octanoic acid

Figure 6: Distribution (upper squares) and enantiomeric compositions (lower squares) of products obtained by CCL-catalyzed reactions of 2- and 3-hydroxyhexanoates

In order to confirm these reversals of the enantioselectivities we investigated analogous CCL-catalyzed reactions of chiral hydroxyacid esters in aqueous medium. Both the hydrolysis of hydroxyacid esters and the cleavage of acetoxyacid esters were characterized by a change of the preferably attacked enantiomer, depending on the position of the asymmetric center.

Similar effects have been observed for the biocatalyzed hydrolysis of racemic methyl branched thiolesters (*29*). The enantiodiscrimination exhibited by CCL and other lipase preparations varies depending on the position of the methyl group.

Based on these results obtained by means of structural variations of the substrates, models of the ester-enzyme complex have been developed (*29*). However, more information of the enzyme region involved will be required, in order to fully understand the mechanisms underlying these phenomena and to predict eventually which enantioselectivity can be expected in the course of these lipase-catalyzed kinetic resolutions.

Conclusions

Lipase from *Candida cylindracea* is a useful biocatalyst to achieve kinetic resolutions of the enantiomers of 2-methyl branched compounds via esterification and transesterification (acidolysis) carried out in organic medium. Optimization of the reaction conditions, e.g. immobilization of the lipase or adjustment of the water content of the solvent, should increase the synthetic value of these processes.

Analogous reactions of chiral hydroxyacid esters turned out to be insignificant in terms of preparative application; however, the study of the pathways involved extended our knowledge on the enzyme-catalyzed formation of optically active acyloxyacid esters. Comparisons of results obtained in simplified model systems with those from complex natural matrices require cautiousness. But the present investigations underline that the unexpected enantiomeric ratios found for hydroxy- and acetoyxacid derivatives in natural systems may not only be rationalized by the action of different enzymes but also by combinations of and/or competition between various enantiodiscriminating pathways.

Considering the high stability of lipases and proteases in organic media a continuing activity of these enzymes after homogenization of plant tissue and subsequent extraction with an organic solvent seems plausible. Therefore further studies should be conducted to find out whether enantiodiscriminating reactions comparable to those described in the present model systems, might also occur during the work-up procedure applied to isolate volatiles from plants, thus altering the "primary" enantiomeric compositions of chiral constituents.

Literature Cited

1.   Schneider, M.P.; Laumen, K. In *Bioflavour '87* ; Schreier, P., Ed.; Walter de Gruyter, Berlin, New York **1988** , pp. 483-529.
2.   Mukherjee, K.D. *Biocatalysis* **1990**, *3*, 277-293.
3.   Zaks, A.; Klibanov, A.M. *Proc. Natl. Acad. Sci., USA.* **1985**, *82*, 3192-3196.

4.   Dordick, J.S. *Enzyme Microb. Technol.* **1989**, *11*, 194-211.
5.   Sih, C.J.; Wu, S.-H. *Top. Stereochem.* **1989**, *19*, 63-125.
6.   Klibanov, A.M. *Acc. Chem. Res.* **1990**, *23*, 114-120.
7.   Gatfield, I.L., *Ann. New York Acad. Sci.* **1984**, *434*, 569-572.
8.   Gillies, B.; Yamazaki, H.; Armstrong D.W., *Biotechnol. Lett.* **1987**, *9*, 709-714.
9.   Triantaphylides, C.; Langrand, G.; Millet, H.; Rangheard, M.S.; Buono, G.; Barrati, J., In *Bioflavour '87*; Schreier, P., Ed.; Walter de Gruyter, Berlin, New York **1988**, 531-542.
10. Gerlach, D.; Lutz, D.; Mey, B.; Schreier, P. *Z. Lebensm. Unters. Forsch.* **1989**, *189*, 141-143.
11. Engel, K.-H., In *Bioflavour '87*; Schreier, P., Ed.; Walter de Gruyter, Berlin, New York **1988**, 75-88.
12. Engel, K.-H.; Heidlas, J.; Albrecht, W.; Tressl, R., In *Flavor Chemistry: Trends and Developments*; Teranishi, R.; Buttery, R.G.; Shahidi, F., Eds.; ACS Symposium Series 388, Washington, D.C. **1989**, 8-22.
13. Flath, R.A.; Black, D.R.; Guadagni, D.G.; McFadden W.H.; Schultz, T.H.; *J. Agric. Food Chem.* **1967**, *15*, 29-34.
14. Idstein, H.; Bauer, C.; Schreier, P. *Z. Lebens. Unters. Forsch.* **1985**, *180*, 394-397.
15. Takeoka, G. R.; Flath, R.A.; Mon, T.R.; Buttery, R.G.; Teranishi, R.; Güntert, M.; Lautamo, R.; Szejtli, J. *J. High Res. Chromatogr.* **1990**, *13*, 202-206.
16. Mosandl, A.; Rettinger, K.; Fischer, K.; Schubert, V.; Schmarr, H.-G.; Maas, B. *J. High Res. Chromatogr.* **1990**, *13*, 382-385.
17. Takeoka, G.R.; Buttery, R.G.; Teranishi, R.; Flath, R.A.; Güntert, M. *J. Agric. Food Chem.* **1991**, *39*, 1848-1851.
18. Sonnet, P.E. *J. Chem. Ecol.* **1984**, *10*, 771-781.
19. Sonnet, P.E. *J. Org. Chem.* **1987**, *52*, 3477-3479.
20. Lutz, D.; Güldner, A.; Thums, R.; Schreier, P. *Tetrahedron Asym.* **1990**, *1*, 783-792.
21. Kirchner, G.; Scollar, M.P.; Klibanov, A.M. *J. Am. Chem. Soc.* **1985**, *107*, 7072-7076.
22. Fukui, T.; Kawamoto, T.; Sonomoto, K.; Tanaka, A. *Appl. Microbiol. Biotechnol.* **1990**, *34*, 330-334.
23. Kodera, Y.; Takahashi., K.; Nishimura, H.; Matsushima, A.; Saito, Y.; Inada, Y. *Biotechnol. Lett.* **1986**, *8*, 881-884.
24. Lazar, G. *Fette, Seifen, Anstrichm.* **1985**, *87*, 394-400.
25. Macrae, A.R. In *Biocatalysis in Organic Syntheses*; Tramper, J., V.D.; Plas, H.C.; Linko, P.; Eds.; Elsevier, Amsterdam, **1985**, 195-208.
26. Yoshimoto, T.; Takahashi, K.; Nishimura, H.; Ajima, A.; Tamaura, Y.; Inada, Y. *Biotechnol. Lett.* **1984**, *6*, 337-340.
27. Engel, K.-H. *Tetrahedron Asym.* **1991**, *2*, 165-168.
28. Engel, K.-H. *J. Am. Oil Chem. Soc.* **1991**, in press.
29. Sonnet, P.E.; Baillargeon, M.W. *Lipids* **1991**, *26*, 295-300.

30. Holmberg, E.; Holquist, M.; Hedenström, E.; Berglund, P.; Norin, T.; Högberg, H.-E.; Hult, K. *Appl. Microbiol. Biotechnol.* **1991**, *35*, 572-578.

31. Chen, C.-S.; Fujimoto, Y.; Girdaukas, G.; Sih, C.J. *J. Am. Chem. Soc.* **1982**, *104*, 7294-7299.

32. Chen, C.-S.; Wu, S.-H.; Girdaukas, G.; Sih, C.J. *J. Am. Chem. Soc.* **1987**, *109*, 2812-2817.

33. Langrand, G.; Baratti, J.; Buono, G.; Trianthaphylides, C. *Biocatalysis* **1988**, *1*, 231-248.

34. Wu, S.-H.; Guo, Z.-W.; Sih, C.J. *J. Am. Chem. Soc.* **1990**, *112*, 1990-1995.

35. Tressl, R.; Engel, K.-H. In *Analysis of Volatiles*, P. Schreier (Ed.); Walter de Gruyter, Berlin, New York, **1984**, 323-342.

36. Tressl, R.; Engel, K.-H. In *Progress in Flavor Research 1984*, J. Adda (Ed.); Elsevier, Amsterdam, **1985**, 441-455.

37. Tressl, R.; Engel, K.-H.; Albrecht, W.; Bille-Abdullah, H. In *Characterization and Measurement of Flavor Compounds*, Bills, D. D.; Mussinan, C.J. (Eds.), ACS Symposium Series 289, Washington D.C., **1985**, 43-60.

38. Engel, K.-H.; Bohnen, M.; Dobe, M. *Enzyme Microb. Techn.* **1991**, *13*, 655-660.

RECEIVED January 13, 1992

Chapter 4

Carboxylester-Lipase-Mediated Reactions
A Versatile Route to Chiral Molecules

Detlef Lutz, Manfred Huffer, Doris Gerlach, and Peter Schreier

Lehrstuhl Für Lebensmittelchemie, Universität Würzburg, Am Hubland, D–8700 Würzburg, Germany

Two classes of naturally occurring flavor compounds were prepared in optically enriched forms by means of lipase catalyzed transformations in organic solvents, i.e. γ- and δ-lactones as well as secondary alcohols. The lactones were obtained *via* intramolecular transesterification reactions, while the chiral secondary alcohols were prepared *via* kinetic resolution using esterification with an achiral acid. For the latter compounds a "model" was developed describing reactivity and selectivity of the enzymatic catalysis, thus providing a powerful tool to establish future synthetic strategies.

Chiral secondary alcohols, esters and lactones are widespread compounds often found in food flavors (1). They comprise saturated, unsaturated and aromatic alcohols, their fatty acid esters, and γ- and δ-lactones, occurring in distinct enantiomeric ratios in nature (2-5). Their production in preparative scale is of interest due to different reasons: (i) If these substances are obtained in optically enriched or pure forms, analytical properties may be evaluated, i.e. enabling us to establish their absolute configurations or to determine the order of elution in chiral chromatography. (ii) After the separation of the enantiomers it is also possible to characterize their selective sensory properties (6). (iii) Finally, "biosynthetic" preparations of industrially attractive compounds with natural enantiomeric ratios are versatile ingredients in food flavoring (7,8).

0097–6156/92/0490–0032$06.00/0
© 1992 American Chemical Society

What are the techniques to get a target molecule highly optically enriched ? Besides chemical methods which are often sophisticated (9), and chromatographic means (10), which are connected with high costs, there is a third way based on biotransformation. Since Klibanov's first report on the catalytic activity of enzymes in organic solvents (11), enzymatic reactions have become more and more important in the synthesis of fine chemicals (12).

How is it possible to obtain optically enriched alcohols, esters and lactones by means of enzyme mediated bioconversions ? There are three general pathways to use a carboxylester lipase system to achieve this aim, i.e. (i) hydrolysis, (ii) esterification, and (iii) transesterification. From the methods shown in Figure 1 we have used both the esterification and transesterification in organic solvents. In the following, the production of γ- and δ-lactones as well as secondary alcohols by carboxylester lipase is described.

Enzyme Selection

First of all, a screening was performed to select the most appropriate carboxylester lipase. A number of preparations are commercially available including catalysts from microbial, plant, and animal sources. They were checked in an esterification system consisting of (R,S)-2-octanol and dodecanoic acid in n-hexane (13). The results are summarized in Figure 2. From the six preparations studied only two provided satisfying findings. The enzyme from *Candida cylindracea* showed a conversion rate of approximately 70 %, but exhibited very low selectivity. As to the selectivity - the prior criterium for our selection - the extract from pig pancreas (PPE) was superior to all other technical enzymes (ee = 85 %). This preparation, which showed a conversion rate of approximately 20 %, was selected for further studies. It should be noted that the (R)-enantiomer was preferably esterified by all the enzyme preparations under study.

Production of Optically Enriched Lactones

The reaction pathway leading to optically enriched lactones is outlined in Figure 3. The interesting compounds, i.e. γ- and δ-lactones with chain lengths from C_5 to C_{13} were chemically converted into the corresponding hydroxypropylesters (14) that were subsequently transformed in organic solvent *via* intramolecular transesterification using PPE. Chiral analysis of the hydroxyesters was performed by their derivatization into diastereomers using chiral agents, such as (R)-(+)-α-methoxy-α-trifluoromethylphenylacetylchloride ((R)-MTPA-Cl), (R)-(+)-α-phenylethylisocyanate ((R)-PEIC), and (S)-O-acetyllactic acid ((S)-ALA-Cl) (15), and by subsequent gas chromatographic

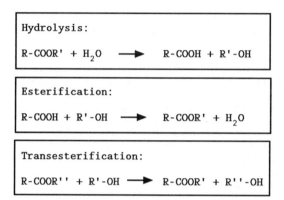

Figure 1.   Scheme of lipase catalyzed reactions.

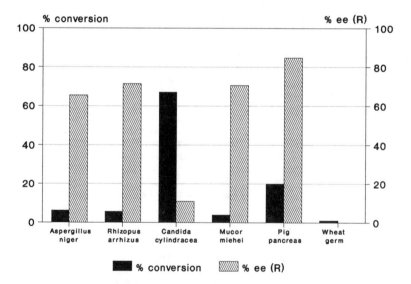

Figure 2.   Screening of lipases by esterification of (R,S)-2-octanol and dodecanoic acid in hexane at 40°C (24 h) (13).

separation (HRGC). Analysis of the obtained lactones was carried out without derivatization by means of high performance liquid chromatography (HPLC) using ChiraSpher (Merck) as chiral stationary phase (10).

The results obtained by the enzymatic transformation of 4-hydroxyesters are illustrated in Figure 4 for the C_5- and C_{11}-homologues. In all cases the resulting lactones were (S)-configurated, whereas the remaining hydroxyesters had (R)-configuration (they could be rearranged chemically to yield the corresponding (R)-lactones).

The following conclusions can be drawn for the PPE mediated production of γ-lactones: (i) high optical yields (ee = > 80 %) were obtained for either lactone or hydroxyester; (ii) moderate reaction rates (10-40 h) were observed, the long-chain homologues reacting faster than the short-chain compounds; (iii) after having reached a maximum value, the ee values of the lactones decreased slightly due to chemical relactonization of the esters.

The results obtained from the PPE catalyzed conversions of 5-hydroxyesters are outlined in Figure 5 for the C_{11} homologue. In comparison to the 4-hydroxyesters the reaction rates determined for the 5-hydroxyesters were five to six times lower. In addition, low enantioselectivity (ee < 20 %) was observed. Therefore, the preparation of optically enriched δ-lactones *via* transesterification of 5-hydroxyesters is strongly limited by the substrate specificity of PPE.

Production of Optically Enriched Alcohols

By reaction of a racemic alcohol with an achiral fatty acid under lipase catalysis the alcohol enantiomers are converted at different reaction rates, resulting in optical enrichment of the produced ester and the remaining alcohol (kinetic resolution, cf. Fig. 6) (16).

In our study to produce relevant flavor alcohols and esters by kinetic resolution, we were successful to establish a "model" describing reactivity and selectivity of the PPE catalyzed esterification in organic media (17). Similar descriptions of enzymic activities have been provided by Ohno et al. (18) using pig liver esterase (PLE) and, recently, by Ehrler and Seebach (19) in their report on the selectivity of PPE in water systems.

First of all, standard experimental conditions were developed to efficiently esterify chiral (R,S)-alcohols. The following reaction conditions were chosen for all transformations: To a solution of 2.5 mmol of each racemic alcohol and dodecanoic acid in 25 ml n-hexane 1 g PPE was added and the mixture stirred at 70°C for 24 hours. After

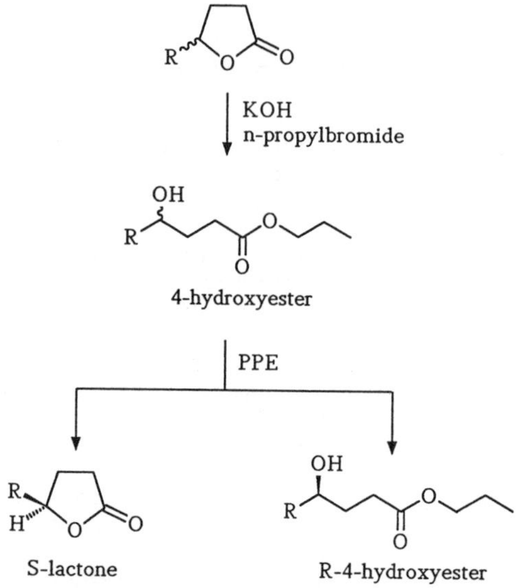

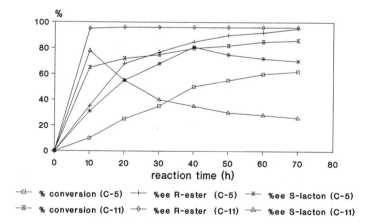

Figure 3. Reaction pathway leading to optically enriched γ-lactones. PPE = porcine pancreas extract (14,15).

Figure 4. PPE catalyzed transesterification of 4-hydroxyesters in diethyl ether at 20°C (15).

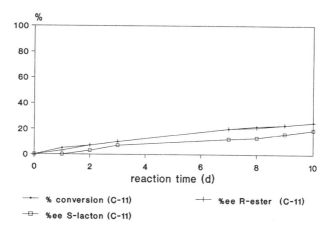

Figure 5. PPE catalyzed transesterification of 5-hydroxyesters in diethyl ether at 20°C (15).

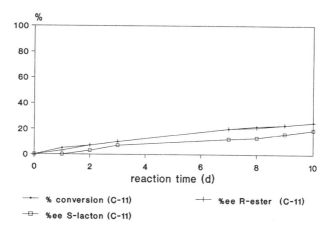

Figure 6. Lipase catalyzed esterification of an (R,S)-alcohol (16).

cooling, the enzyme was filtered off and the reaction
mixture worked up for the analysis of products and educts.
Esters were reductively cleaved yielding the original
alcohols which could be easily analyzed using the same
derivatization agents as mentioned previously for the hydroxyesters or using the chiral Lipodex C column without
any derivatization (17). Configurations were determined
with the help of polarimetry, in case literature data
existed, or by NMR experiments, if those were lacking
(20). For more detailed analytical information see reference (17).

a. *2-Alkanols and E-3-Alken-2-ols*

The results obtained in the homologous series of 2-alka-
nols and E-3-alken-2-ols are outlined in Table I. The conversion rates for the saturated substrates ranged from
20-30 %, whereas those of the unsaturated analogs were
about 40 %. It seemed that the alkenols were activated by
the influence of the double bond. As to the selectivity,
there was no significant difference between the two classes. In all cases, the preferred esterified alcohol was
(R)-configurated with considerably high ee values (about
90 %).

Table I. PPE catalyzed esterification of 2-alkanols and
E-3-alken-2-ols

Substrate	% Ester	ee Ester	ee Alcohol	Preferred Antipode
2-Pentanol	21.2	87.4	26.8	R
2-Hexanol	30.1	95.9	33.8	R
2-Heptanol	27.0	92.3	33.9	R
2-Octanol	31.3	93.1	41.4	R
2-Nonanol	19.3	92.7	25.1	R
E-3-Penten-2-ol	43.0	79.1	77.6	R
E-3-Hepten-2-ol	40.3	86.1	97.0	R
E-3-Octen-2-ol	42.6	93.8	97.7	R

b. *1-Alken-3-ols*

Conversion rates for the 1-alken-3-ols ranged from 15-30
%. The results obtained in this series concerning selectivity are illustrated in Figure 7. As shown from the
graph, the configuration of the preferably converted enantiomers changed between the C_5 and the C_6 alkenol. Such a
reversal of configuration has been previously observed in
PLE catalyzed hydrolyses (21). For better understanding
this phenomenon, the structures of the preferred enantiomers are outlined as follows:

In this projection, the shorter, less bulky side chains

are in the top position, whereas the longer (larger) side chains are at the bottom. In the homologous series the reversal of configuration occurred when both side chains were sterically equal, i.e. in the case of 1-penten-3-ol. At the same time the absolute ee value reached a minimum. The preliminary "model" of the reactive enantiomers as described above should be approved in the following studies.

c. *1-Alken-4-ols*

Figure 8 shows the results of the conversions obtained in the series of 1-alken-4-ols. Again a reversal of the configuration was observed when the side chains became sterically equal, i.e. between the C_6 and the C_7 homologue. The reaction rates decreased from 36 % (1-penten-4-ol) to 20 % (1-octen-4-ol), an effect that became even more drastic in the series of E-2-alken-4-ols.

d. *E-2-Alken-4-ols*

In this series of experiments, a decrease of ester formation rate with increasing chain length was observed (Table II). The reduced accessibility of the reactive hydroxy group, which is situated in the middle of the molecules in the case of the higher homologues, may be responsible for this effect. In analogy to the 1-alken-3-ols and 1-alken-4-ols a reversal of configuration of the preferably converted enantiomers was found, again at the position where both side chains were sterically equal, i.e. between C_7 and C_8.

e. *Isomeric Hexenols*

Not only in homologous series, but also in the case of isomeric compounds the phenomenon of reversal of configuration was observed as illustrated in Figure 9. Due to the previously mentioned reason, i.e. the accessibility of the reactive group, 1-hexen-5-ol was found to be the most reactive racemate in this series (conversion rate, 43 %).

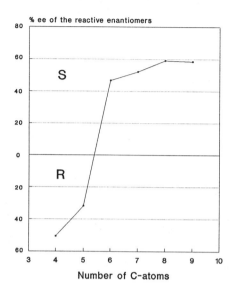

Figure 7. PPE catalyzed esterification of homologous 1-alken-3-ols. (Reproduced with permission from reference 17. Copyright 1990 Pergamon.)

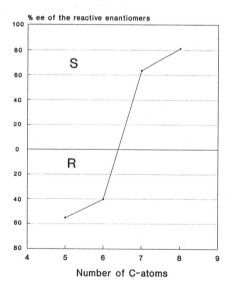

Figure 8. PPE catalyzed esterification of homologous 1-alken-4-ols. (Reproduced with permission from reference 17. Copyright 1990 Pergamon.)

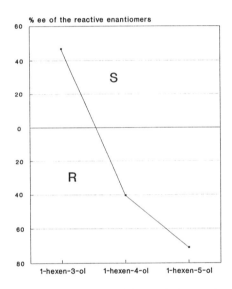

Figure 9. PPE catalyzed esterification of isomeric hexenols. (Reproduced with permission from reference 17. Copyright 1990 Pergamon.)

Table II. PPE catalyzed esterification of E-2-alken-4-ols

Substrate	% Ester	ee Ester	ee Alcohol	Preferred Antipode
E-2-Penten-4-ol	43.0	79.1	77.6	R
E-2-Hexen-4-ol	33.7	90.0	60.8	R
E-2-Hepten-4-ol	12.0	71.9	11.1	R
E-2-Octen-4-ol	5.7	80.0	1.1	S

f. *Z/E-Isomeric Alkenols*

The ester formation rates of Z-isomers were determined to
be considerably lower than those of the analogous E-isomers (Table III). Obviously, the reactive center is pro-
tected by the particular geometry of the allylic Z-double
bond. In all cases, the (R)-enantiomer was the preferred
substrate. However, the esterification of E-isomers exhibited much higher selectivity which is also an effect of
substrate and enzyme geometry.

Table III. PPE catalyzed esterification of Z- and E-confi-
gurated alkenols

Substrate	% Ester	ee Ester	ee Alcohol	Preferred Antipode
Z-3-Hepten-2-ol	6.9	53.3	1.2	R
E-3-Hepten-2-ol	40.3	86.1	97.0	R
Z-3-Octen-2-ol	14.3	40.6	9.1	R
E-3-Octen-2-ol	42.3	93.8	97.7	R

g. *Aromatic Alcohols*

In this class of compounds nearly constant and high ee
values (about 90 %, cf. Table IV) were found. Obviously,
the aromatic rings in the substrates are the main factors
determining the selectivity. In the above illustrated
stereochemical projection (cf. b.) the large benzene ring
would be at the bottom position, i.e. also the aromatic
substrates are in accordance with the postulated "model".
The conversion rates measured were dependent on the position of the hydroxy group (C_2 or C_3) and ranged from 20 to
40 %.

Table IV. PPE catalyzed esterification of aromatic alcohols

Substrate	% Ester	ee Ester	ee Alcohol	Preferred Antipode
1-Phenyl-1-ethanol	33.0	92.7	52.7	R
1-Phenyl-2-propen-1-ol	19.3	85.7	30.0	R
1-Phenyl-1-propanol	18.0	86.9	19.6	R
4-Phenyl-E-3-buten-2-ol	43.8	88.5	73.7	R
4-Phenyl-2-butanol	28.9	83.3	44.4	R

Conclusions

As a general result it can be stressed that the preferably esterified enantiomers exhibit the following configurations:

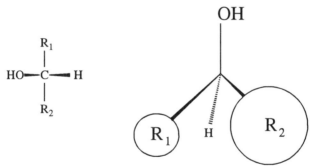

R_1 is the shorter, less bulky, R_2 the longer or larger side chain.

The kinetic resolution of a racemate is a two substrate reaction. If we look at the <u>reactivity</u> of a single enantiomer, it can be ruled out that a short R_1 will result in high reactivity; in the optimal case R_1 is a methyl group. Therefore, the most reactive compounds are ...2-ols. Conversely, large R_1, like a benzene ring, will lead to low reactivity. This may be illustrated by two examples:

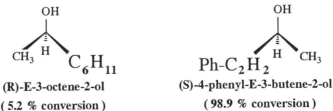

(R)-E-3-octene-2-ol

(5.2 % conversion)

(S)-4-phenyl-E-butene-2-ol

(98.9 % conversion)

A racemate is transformed with high <u>enantioselectivity</u>, if the two enantiomers differ strongly in their reactivities, i.e. the two side chains have to be fairly different in their sterical dimensions. A good example for this statement is (R,S)-2-octanol which is esterified with an enantioselectivity of 93 % ee.

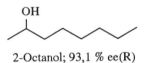

2-Octanol; 93,1 % ee(R)

However, some inconsistencies were found, i.e. the relatively high ee values determined for 1-hepten-3-ol. But it should be considered that the studies were carried out with a crude protein extract. In order to characterize the active enzyme(s) according to its (their) kinetic and structural features, purification is needed. Actually, there are some hints that the common triacylglyceride lipase of pig pancreas is not responsible for the catalysis in organic media (22,23). Work to characterize the effective enzyme is in progress in our laboratory.

Acknowledgement. The financial support of this work by the Deutsche Forschungsgemeinschaft, Bonn, is gratefully acknowledged. The authors thank A. Güldner and R. Thums for their helpful collaboration.

LITERATURE CITED

1. Bauer, K.; Garbe, D.; Surburg, H. *Common Fragrance and Flavor Materials,* 2nd. ed., VCH Verlagsgesellschaft, Weinheim, 1990.
2. Engel, K.H. In: *Bioflavour '87*, Schreier, P., ed.; W. de Gruyter: Berlin, New York, 1988, pp. 89-95.
3. Fröhlich, O.; Huffer, M.; Schreier, P. Z. Naturforsch. **1989,** *44c,* 555.
4. Mosandl, A. *Food Rev. Intern.* **1988,** *4,* 1.
5. Bernreuther, A.; Christoph, N.; Schreier, P. *J. Chromatogr.* **1989,** *481,* 363.
6. Mosandl, A.; Heusinger, G.; Gessner, M. *J. Agric. Food Chem.* **1986,** *34,* 119.
7. Drawert, F. In: *Bioflavour '87*, Schreier, P., ed., W. de Gruyter: Berlin, New York, 1988, pp. 3-32.
8. Schreier, P. *Food Rev. Intern.* **1989,** *5,* 289.
9. Helmchen, G.; Ihrig, K.; Schindler, H. *Tetrahedron Lett.* **1987,** *28,* 183.
10. Huffer, M.; Schreier, P. *J. Chromatogr.* **1989,** *469,* 137.

11. Cambou, B.; Klibanov, A.M. *J. Am. Chem. Soc.* **1984,** *106,* 2687.
12. Crout, D.H.G.; Christen, M. In: *Modern Synthetic Methods 1989,* Scheffold, C.R., ed., Springer: Berlin, Heidelberg, New York, pp. 1-114.
13. Gerlach, D. Doctoral Thesis, University of Würzburg, 1988.
14. Mosandl, A.; Günther, C. *J. Agric. Food Chem.* **1989,** *37,* 413.
15. Huffer, M. Doctoral THesis, University of Würzburg, 1990.
16. Chen, C.S.; Fujimoto, Y.; Girdaukas, G.; Sih, C.J. *J. Am. Chem. Soc.* **1982,** *104,* 7294.
17. Lutz, D.; Güldner, A.; Thums, R.; Schreier, P. *Tetrahedron Asymmetry* **1990,** *1,* 783.
18. Ohno, M.; Kobayashi, S.; Adachi, K. In: *Enzymes as Catalysts in Organic Synthesis,* Schneider, M.P., ed.; Reidel: Dordrecht, 1986, pp. 123-142.
19. Ehrler, J.; Seebach, D. *Liebigs Ann. Chem.* **1990,** 379.
20. Helmchen, G.; Schmierer, R. *Angew. Chem.* **1976,** *13,* 770.
21. Björkling, F.; Boutelje, J.; Gatenbeck, S.; Hult, K.; Norin, T.; Szmulik, P. *Tetrahedron* **1985,** *41,* 1347.
22. Tombo, G.M.R.; Schaer, H.P.; Fernandez I Busquets, X.; Ghisalba, O. *Tetrahedron Lett.* **1986,** *27,* 5707.
23. Hemmerle, H.; Gais, H.J. *Tetrahedron Lett.* **1988,** *28,* 3471.

RECEIVED December 18, 1991

Chapter 5

Biosynthesis and Biotechnological Production of Aliphatic γ- and δ-Lactones

W. Albrecht, J. Heidlas, M. Schwarz, and R. Tressl

Institut für Biotechnologie, Technische Universität Berlin, Fachgebiet
Chemisch-Technische Analyse, Seestrass 13, D–1000 Berlin 65, Germany

The enantiomeric composition of γ- and δ-lactones isolated from peaches was determined. Post-harvest ripening of the fruits was characterized by an increase in the concentration of lactones. The configuration and the optical purity both remained constant.

Biosynthetic studies on γ- and δ-lactones were performed with the lactone producing yeast *Sporobolomyces odorus*. After administration of [9,10,12,13-^{2}H$_4$]-linoleic acid to intact cells of *Sp. odorus*, deuterium labelled (Z)-6-dodecen-4-olide, γ–nonalactone, and δ-decalactone could be detected. The conversion of (S)-13-hydroxy-(Z,E)-9,11-octadecadienoic acid into optically pure (R)-δ-decalactone was elucidated. Based on the results of incubation experiments with deuterium labelled hydroxy- and oxo acids, an oxidation of the secondary hydroxy group followed by a subsequent enantioselective reduction was found to be responsible for the inversion of configuration.

Administered [2,2-^{2}H$_2$]-(E)-3-decenoic acid and [2,2-^{2}H$_2$]-(E)-3,4-epoxydecanoic acid were converted into deuterium labelled γ-decalactone. Based on the abstraction of one ^{2}H atom in the course of the biotransformations and the identification of labelled 2-decen-4-olide, a biosynthetic pathway leading to γ-decalactone is presented.

The biotechnological production of optically pure (R)-γ-decalactone and (R)-(Z)-7-decen-4-olide from ricineloic acid and densipolic acid, respectively, is presented. With chemically modified and deuterium labelled precursors mechanistic and stereochemical features are depicted.

Chiral aliphatic γ- and δ-lactones are widely distributed in nature. As volatiles in several fruits, e.g. peaches, apricots, nectarines, and strawberries they contribute to the overall flavor in a decisive way (1). Due to these properties lactones play an important role in the flavor industry as aroma compounds.

0097–6156/92/0490–0046$06.00/0
© 1992 American Chemical Society

In recent years several gas chromatographic methods for the determination of the configuration and the optical purity of naturally occurring lactones have been developed. For the enantioseparation on non-chiral stationary phases various optically pure reagents were used to convert the lactones into diastereomeric derivatives (2-4). Today the availability of capillary columns coated with various derivatives of α-,β-, and γ-cyclodextrins permit an enantioseparation of γ- and δ-lactones without prior derivatization (5,6). Numerous applications of this convenient technique have been published which describe the chirospecific analysis of lactones isolated from natural sources. According to these data, it is evident that naturally occurring lactones are optically active and both the configuration and optical purity can vary for identical substances isolated from different sources (7,8). The chirospecific analysis of lactones is an acknowledged method to prove the authenticity of natural flavorings in food, although the biosynthesis and thus the formation of distinct enantiomeric ratios is still unknown.

This article summarizes our recent investigations on the biosynthesis of γ- and δ-lactones. A discussion of biosynthetic pathways is given based on the results of incubation experiments with labelled precursors. In addition, strategies for the biotechnological production of optically active lactones via biotransformations of long chain hydroxy acids which were isolated from natural sources are presented.

Variation of the enantiomeric composition of lactones in peaches during post-harvest ripening

Differences in the enantiomeric composition of identical lactones, even determined in the same kind of fruit, led to questions concerning the factors upon which these ratios depend. Will the enantiomeric ratio of lactones change with increasing concentration in the course of ripening or is there a constant value which is characteristic for the variety of the plants? To determine the variations of the enantiomeric composition of lactones in peaches, fresh fruits were stored for six days in the dark at 20°C. During this period the color of the fruits changed from yellow-green to red-yellow and the tissue became soft and juicy. As outlined in Figure 1 the quantitative determination of the volatiles showed a decrease in the concentration of the 'green' aroma components hexanal and (E)-2-hexenal. Simultaneously, an increase in the concentration of all lactones could be observed. However, according to the chirospecific analysis, the enantiomeric ratios of the lactones remained constant (Table 1).

The invariant optical purities can be interpreted in two ways. Firstly, the formation of the lactones could be controlled by more than one pathway with opposite enantioselectivities. However, in this case the activity of the biosynthetic enzymes must remain constant during ripening of the fruit in order to give rise to constant enantiomeric ratios of lactones. Alternatively, a single pathway to the γ-lactones as well as the δ-lactones could be proposed. In this case, the reaction determining the configuration would be catalyzed by enzymes with variable enantioselectivities influenced by the structure and/or the chain length of the precursor.

Incubation Experiments for the Investigation of the Biosynthesis of Lactones

For the investigation of biosynthetic pathways of flavors radiolabelled compounds are generally administered to fruit tissue or enzyme preparations. The transformations of the precursor can then be followed by appropriate radiochemical methods. Schreier reviewed several biosynthetic studies in which the metabolism of ^{14}C labelled precursors were investigated by radio gas chromatography (9).

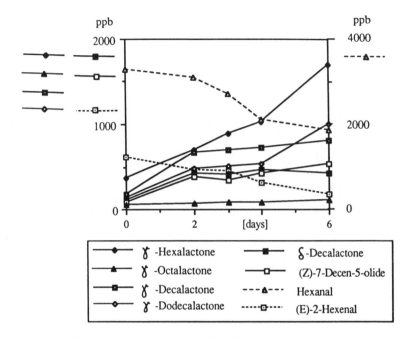

Figure 1. Concentration of volatiles isolated from peaches during post-harvest ripening.

Table 1: Enantiomeric composition of γ- and δ-lactones isolated from peaches during post-harvest ripening

Compound	Enantiomeric composition (R : S [%])			
	0 days	3 days	4 days	6 days
γ-hexalactone	78 : 22	82 : 18	85 : 15	84 : 16
γ-octalactone	88 : 12	89 : 11	86 : 14	n.d.[*]
γ-decalactone	89 :11	89 : 11	90 : 10	90 : 10
γ-dodecalactone	> 98: < 2	> 98: < 2	> 98: < 2	n.d.[*]
δ-decalactone	98 : 2	97 : 3	98 : 2	n.d.[*]
(Z)-7-decen-5-olide	35 : 65	34 : 66	36 : 64	33 : 67

[*] n.d. = not determined

Due to their high sensitivity mass spectrometers and high field NMR-spectrometers can also be applied for biosynthetic studies using precursors which are labelled with either ^2H, ^{13}C, or ^{18}O. In previous reports some results from investigations of the biosynthetic pathways of chiral hydroxy acid esters and lactones which used deuterium labelled precursors were summarized (10,11). For example, the capacity of pineapple slices to transform both, 3-hydroxyhexanoic acid and (Z)-4-octenoic acid into δ-octalactone could be demonstrated. Unfortunately, low biosynthetic activity prevented a chirospecific analysis of the labelled product in order to determine the stereospecificity of the detected reactions.

The general disadvantages of the use of fruit slices for biosynthetic studies are the dependence of the metabolic activity on the variety of the fruit and the physiological condition, i.e. the stage of ripeness. To overcome these problems the carotenoid producing yeast *Sporobolomyces odorus* was used for further investigations; this yeast excretes a series of γ- and δ-lactones during growth (12,13).

The main products which accumulate in the fermentation broth are (R)-(Z)-6-dodecen-4-olide (92 % ee) and (R)-γ-decalactone (>98 % ee). In shake flasks concentrations of 100 ppm and 20 ppm, respectively, could be isolated after a fermentation time of 10 days. Maximum concentrations of 5 ppm (R)-δ-decalactone and (S)-(Z)-7-decen-5-olide accumulated after 3-4 days. In the course of the cultivation, biosynthesized δ-lactones are reabsorbed by the organism whereas the γ-lactones represent metabolic end products. According to the quantitative determination of volatiles in the fermentation broth of *Sp. odorus* during cultivation in a 10 L fermenter, the γ-lactones are mainly formed during a period when previously biosynthesized long chain fatty acids are degraded (10). Therefore, a relationship between the β-oxidation and the biosynthesis of γ-lactones was assumed.

In the course of the β-oxidation of fatty acids, unsaturated derivatives were formed. An intermediate in the degradation of linoleic acid is (E)-3-decenoyl-CoA, generated by a NADPH-dependent reduction of (E,Z)-2,4-decadienoyl-CoA (14). The formation of γ-decalactone from (E)-3-decenoic acid and from (E)-3,4-epoxydecanoyl-CoA could be demonstrated with deuterium labelled precursors (15,16). The transformation of $[2,2-^2H_2]$-(E)-3-decenoic acid was accompanied by an abstraction of one deuterium. A very low incorporation, however, prevented the determination of the stereospecificity of this trans-formation. The analysis of the incubation experiment with racemic $[2,2-^2H_2]$-(E)-3,4-epoxydecanoic acid is summarized in Figure 2. As in the case of (E)-3-decenoic acid, the GC-MS analysis of γ-decalactone extracted after incubation of the epoxide revealed the loss of one deuterium atom in the course of the biotransformation. In order to determine the configuration and the optical

purity of γ-decalactone, the compound was isolated by preparative GC and transformed into diastereomeric 4-[(R)-[(1-Phenylethyl)-carbamoyl]oxy]-N-butyldecan-amides (3). In contrast to the high enantiomeric excess of the biosynthesized lactone, the product of the incubation experiment was a mixture of both enantiomers. Several repetitions of these experiments revealed that the enantiomeric composition is dependent on the ratio of biosynthesized lactone to transformed precursor.

To estimate the portion of labelled lactone in both enantiomers the diastereo-isomers were investigated by GC-MS. The mass spectra of the derivatives are characterized by the removal of the phenylethylcarbamate moiety leading to the fragment ion [M-164]⁺. The presence of the ion m/z 227 showed that the (S)-enantiomer had originated from the racemic precursor. The (R)-enantiomer consisted of the biosynthesized lactone (m/z 226) and the transformed epoxide (m/z 227). Thus, the determination of the ratio of labelled

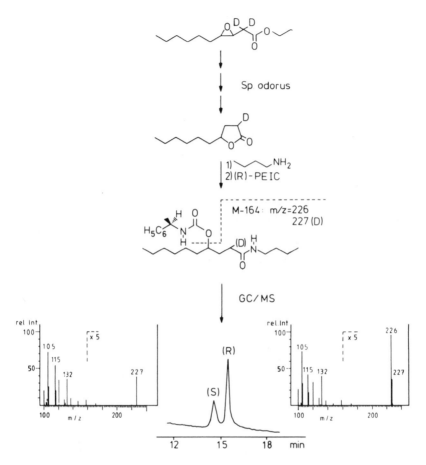

Figure 2. Biotransformation of racemic [2,2-2H2]-(E)-3,4-epoxydecanoic acid into γ-decalactone by *Sporobolomyces odorus*.

to genuine lactone in addition to the enantioseparation revealed a non specific metabolism of racemic (E)-3,4-epoxydecanoic acid.

Furthermore, the precursor was transformed into labelled 2-decen-4-olide (16). This compound does not accumulate in cultures of *Sp. odorus* unless the substrate is added to the incubation mixture. 4-Hydroxy-(Z)-2-decenoic acid or its Coenzyme A derivative can be assumed as the immediate precursor of the unsaturated lactone. Based on these results a biosynthetic pathway for γ-decalactone was proposed (Figure 3). (E)-3-Decenoyl-CoA, the product from the β-oxidation of linoleic acid, is enantioselectively epoxidized to (3R,4R)-(E)-3,4-epoxydecanoyl-CoA. Hydration of the epoxide with subsequent dehydration and reduction leads to (R)-4-hydroxy-decanoyl-CoA, and after cyclization, to γ-

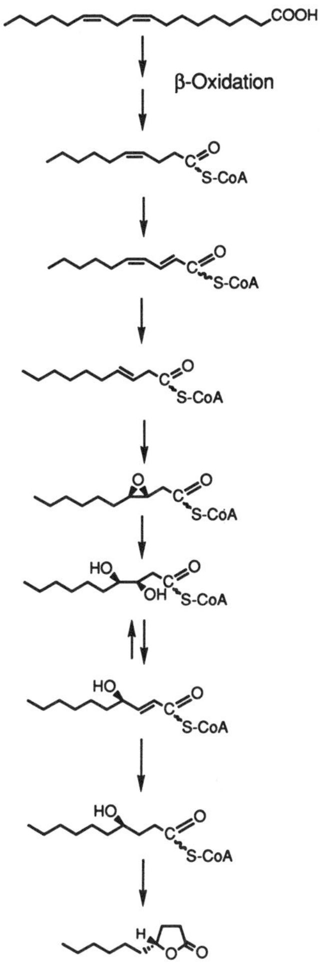

Figure 3. Proposed biosynthetic pathway of γ-decalactone in *Sporobolomyces odorus*.

decalactone. Additionally, this pathway provides an explanation for the formation of 2-alken-4-olides along with the corresponding saturated γ-lactones by the basidiomycete *Polyporus durus* (17).

To prove the proposed mechanism [9,10,12,13-^{2}H$_4$]-linoleic acid was fed to intact cells of *Sp. odorus*. This precursor was synthesized by the route shown in Scheme 1. 2-Octynebromide and 9-decynoic acid were coupled yielding 9,12-octadecadiynoic acid. Finally, the acetylenic bonds were then

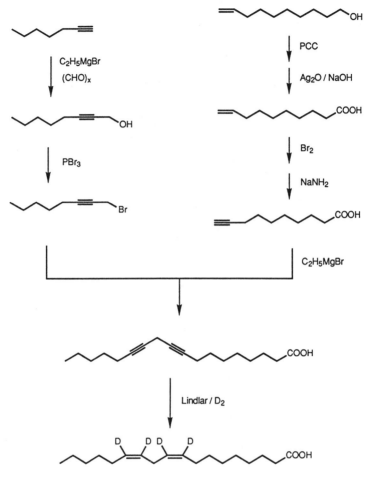

Scheme 1. Synthesis of [9,10,12,13-^{2}H$_4$]-linoleic acid.

partially reduced with ^{2}H$_2$ in the presence of a Lindlar catalyst (J. Heidlas, R. Tressl, unpublished data).

After administration of [9,10,12,13-^{2}H$_4$]-linoleic acid to eight day old cultures a transformation into (Z)-6-dodecen-4-olide and γ-nonalactone could be detected. γ-Decalactone, however, contained no deuterium. The mass spectra of the labelled and the genuine compounds are shown in Fig. 4. The molecular peak at m/z 200 showed that the main product, (Z)-6-dodecen-4-olide possessed all the four deuterium atoms of the precursor. In addition, the fragment ion for the lactone moiety at m/z 87 revealed the presence of two deuterium atoms attached to the ring. No intermediates on the pathway leading to this lactone could be detected thus preventing a more detailed description of the biosynthetic pathway. The identification of labelled γ-nonalactone and the formation of 9-oxononanoic acid, which accumulated in minor amounts indicated the hydroperoxidation of linoleic acid. Therefore a stereospecific lipoxygenase mediated

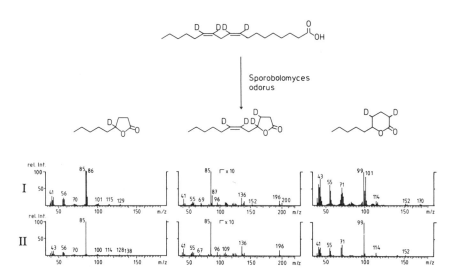

Figure 4. Mass spectra of lactones isolated after incubation of [9,10,12,13-^{2}H$_4$]-linoleic acid to cultures of *Sporobolomyces odorus*; (I: deuterium labelled compounds; II: genuine reference compounds).

reaction leading to (S)-10-hydroperoxy-(E,Z)-8,12-octadecadienoic acid can be assumed. In edible mushrooms and various microorganisms this reaction initiates the formation of (R)-1-octene-3-ol and 10-oxo-(E)-8-decenoic acid (18,19). Since this organism appears to lack a lyase catalyzing the cleavage of the hydroperoxide, a reduction, leading to the corresponding optically active hydroxy acid and subsequent β-oxidation seems probable for the biosynthesis of (R)-(Z)-6-dodecen-4-olide.

When [9,10,12,13-^{2}H$_4$]-linoleic acid was added to cultures of *Sp. odorus* active in the biosynthesis of δ-lactones a transformation into δ-decalactone could be observed. According to the mass spectrum shown in Figure 4 the lactone possessed two deuterium atoms (m/z 101, 116, 172 (M$^+$)) indicating that a deuteron must have been removed in addition to that located at C-9, which represents the carboxyl carbon of the lactone. The hydroxy function must be incorporated at the ω-6 carbon. With ^{18}O labelled (S)-13-hydroperoxy- and (S)-13-hydroxy-(Z,E)-9,11-octadecadienoic acid (coriolic acid), prepared from linoleic acid by soybean lipoxygenase catalyzed hydroperoxidation under ^{18}O$_2$ atmosphere, the capacity of *Sp. odorus* to convert both precursors into δ-decalactone was demonstrated (W. Albrecht, R. Tressl, unpublished data). As determined with the unlabeled hydroxy acid, added after the degradation of previously biosynthesized δ-decalactone, the biotransformation led to optically pure (R)-δ-decalactone. Furthermore, 13-oxo-(Z,E)-9,11-octadecadienoic acid could also be identified as a product of this biotransformation indicating an inversion of configuration of the precursor via an oxidation and a subsequent enantioselective reduction. To confirm this proposed mechanism racemic [13-^{2}H]-13-hydroxy-, (S)-[9,10,12,13-^{2}H$_4$]-13-hydroxy-, and 13-oxo-(Z,E)-9,11-octadecadienoic acid were used as substrates. The results of these biotransformations are summarized in Table 2.

Table 2: Biotransformations of hydroxy- and oxoacids into optically pure (R)-δ-decalactone catalyzed by *Sp. odorus*

Precursor	(Configuration)-Product	Enantiomeric purity [% ee]	Conversion [%]
(S)-13-Hydroxy-(Z,E)-9,11-octadecadienoic acid	(R)-δ-Decalactone	> 98	15.0
[13-^{2}H]-13-Hydroxy-(Z,E)-9,11-octadecadienoic acid	(R)-δ-Decalactone	> 98	14.0
13-Oxo-(Z,E)-9,11-octadecadienoic acid	(R)-δ-Decalactone	> 98	10.8
[9,10,12,13-^{2}H$_4$]-(S)-13-Hydroxy-(Z,E)-9,11-octadecadienoic acid	[2,4-^{2}H$_2$]-(R)-δ-Decalactone	> 98	18.7
[9,10,11,12-^{2}H$_4$]-(S)-13-Hydroxystearic acid ethyl ester	[2,3,4-^{2}H$_3$]-(R)-δ-Decalactone	> 98	1.3
[9,10,11,12-^{2}H$_4$]-13-Oxostearic ethyl ester	[2,3,4-^{2}H$_3$]-(R)-δ-Decalactone	> 98	1.2

All precursors were converted into optically pure (R)-δ-decalactone in comparable yields (10.8 - 18.7 %). The lack of deuterium in δ-decalactone generated from the racemic deuterated hydroxy acid indicates that both enantiomers of this precursor must have been oxidized. The lactone generated from (S)-[9,10,12,13-^{2}H$_4$]-13-hydroxy-(Z,E)-9,11-octadecadienoic acid possessed only two deuterium atoms and the mass spectrum was identical to that of δ-decalactone detected after incubation of [9,10,12,13-^{2}H$_4$]-linoleic acid. Based on these results a preliminary biosynthetic pathway can be proposed (Figure 5). After hydroperoxidation of linoleic acid, 13-oxo-(Z,E)-9,11-octadecadienoic acid may be formed either directly or via the hydroxy acid. The generation of the oxo acid in the presence of metal containing proteins was demonstrated by Hamberg (20). After an enantioselective reduction, the (R)-hydroxy acid is degraded to (R)-δ-decalactone. As outlined in Figure 5, the removal of the diene structure requires not only an isomerization of a (Z)-3-enoyl-CoA into a (E)-2-enoyl-CoA but also a reduction of a Δ^3 double bond in 5-hydroxy-(E)-3-decenoyl-CoA, which does not represent a common step in the β-oxidation of fatty acids. To investigate the influence of double bonds on the formation of δ-decalactone, [9,10,11,12-^{2}H$_4$]-13-hydroxy- and -13-oxostearic acid were used as precursors (Table 2). The high optical purity of (R)-δ-decalactone isolated after incubation of both substrates demonstrated that the inversion of the configuration was not dependent on the presence of the diene structure. However, the yield of the product which could be isolated from the fermentation broth decreased dramatically. These results led to the conclusion that the formation of unsaturated intermediates, which do not represent substrates for the β-oxidation enzymes represents a bottleneck in the complete degradation of unsaturated hydroxy acids and thus leads to the temporary accumulation of δ-decalactone in the culture broth of *Sp. odorus*.

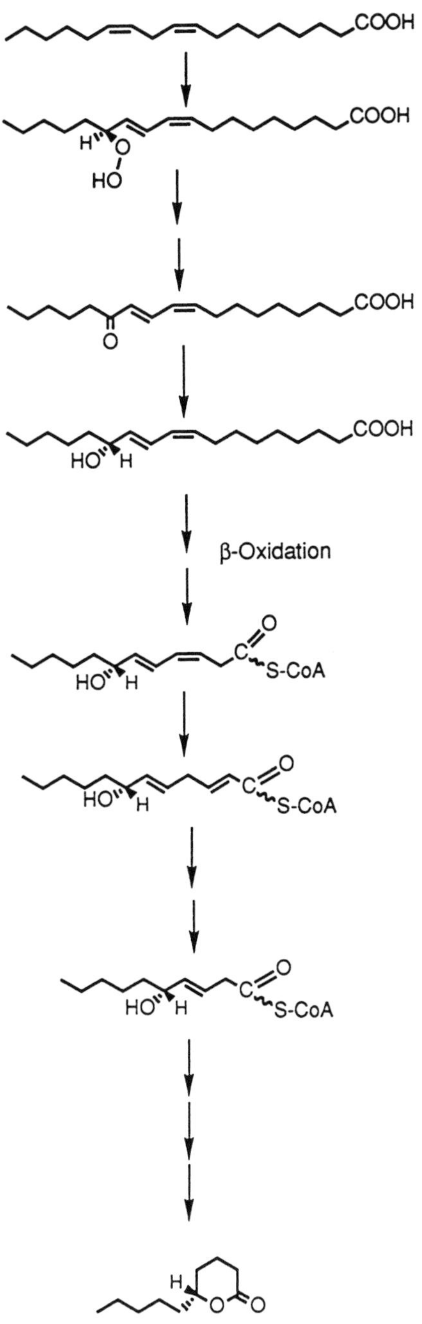

Figure 5. Proposed biosynthetic pathway of δ-decalactone in *Sporobolomyces odorus*.

(16) Albrecht, W.; Tressl, R.; Z. *Naturforsch.* **1990**, *45c*, 207.
(17) Berger, R.G.; Neuhäuser, K.; Drawert, F.; Z. *Naturforsch.* **1986**, *41c*, 963.
(18) Wurzenberger, M.; Grosch, W.; *Biochem. Biophys. Acta*, **1984**, *794*, 25.
(19) Tressl, R.; Bahri, D.; Engel, K.-H.; *J. Agric. Food Chem.* **1982**, *32*, 89.
(20) Hamberg, M.; *Lipids* **1975**, *10*, 87.
(21) Gatfield, I.L.; in Biogeneration of Aromas; Parliment, T.H.; Croteau, R.; Eds.;   ACS Symp. Series 317, 1986, 310 - 319.
(22) US Patent 4,560,656; 1985.
(23) Cardillo, R.; Fronza, G.; Fuganti, C.; Graselli, P.; Nepoti, V.; Barbeni, M; Guarda, P.A.; *J. Org. Chem.* **1989**, *54*, 4979.
(24) Cardillo, R.; Fronza, G.; Fuganti, C.; Grasselli, P.; Mele, A.; Pizzi, D.; Allegrone, G.; Barbeni, M.; Pisciotta, A.; *J. Org Chem.* **1991**, 56, 5237.
(25) Gunstone, F.D.; Harwood, J.L.; Padley, F.B.; Eds.; The Lipid Handbook; Chapman and Hall, London - New York, 1986.
(26) Van der Gen, A.; *Parfumes, Cosmétiques, Savons de France* **1972**, *2*, 356.

RECEIVED January 17, 1992

Chapter 6

Precursor Atmosphere Technology
Efficient Aroma Enrichment in Fruit Cells

Ralf G. Berger[1], Gerd R. Dettweiler[2], Gabriele M. R. Krempler[2], and Friedrich Drawert[2]

[1]Institüt für Lebensmittelchemie der Universität Hannover, Wunstorferstrass 14, D–3000 Hannover 91, Germany
[2]Institut für Lebensmittelchemie und Analytische Chemie der Technische Universität München, D–8050 Freising 12, Germany

Precursor atmosphere (PA)-technology is a short-time storage biotechnology that uses intact, mature fruit cells as a biocatalyst for the production of fruit flavor. Precursor substrate is supplied by exposing fruit tissue to a controlled atmosphere containing vapor of, e.g., volatile alcohols. Increasing the substrate concentration subsequently increased the concentration of certain carboxylic esters in apple cvs.; accumulation factors for ethyl 2-methylbutanoate of 20 in 48 h, and more than 50 in 8 d were recorded. The newly formed esters can be isolated from the PA by adsorption/solvent desorption, or can be transfered into processed products. Analytical and sensory data on pilot scale HTST-juices of PA-stored apple demonstrate a significant improvement of the product quality.

Volatile carboxylic esters constitute an important group of aroma substances in many fruits, e.g., in apple, pear, and banana. The producing cells are distinguished not only by their ability to form, at or beyond the climacteric maximum, esters from substrates not regularly present in healthy plant cells; they are also able to take up and metabolise exogenous substrates when tissue preparations are incubated in a buffer solution according to the "Aged Tissue model" (1). Though in these tissue preparations an indirect stimulation of the flavor metabolism cannot be ruled out completely, the patterns of distribution of volatiles after exposure, as compared to untreated controls, suggest a direct bioconversion and dynamic incorporation into the volatile flavor fraction (2). The operation of several pathways in ripe fruits was

0097–6156/92/0490–0059$06.00/0
© 1992 American Chemical Society

Table I. Concentration of Volatiles after Pressure Injection   of   n-Butanol   into   Apples   cv.   *Jonathan*

Compound (μg/100g)	Control 0 d	Precursor injected[*] 1 d ----- 3 d	
n-Butanol	186	4030	3610
Et butanoate	63	85	98
Bu acetate	100	120	103
Bu butanoate	55	52	91
Bu 2-mebutanoate	17	24	80
Bu hexanoate	67	76	193
He acetate	148	153	213
He butanoate	42	53	107
He 2-mebutanoate	121	164	330

[*] 10 x 500 μg n-Butanol/Fruit

two orders of magnitude higher in the peel section than in the parenchymatic cells of the flesh (*16*). Previous incubation experiments using precursors of esters with separated peel or peel discs (*11,15,17*) also indicated that an acyl-CoA-alkyl-transferase is located subepidermally. A diffusional transport brings aroma compounds to inner parts of the fruit. As expected, the more polar compounds show a less pronounced concentration gradient in the aqueous matrix of the fruit (*16*). With respect to the morphological situation, an administration of volatile precursors of aroma esters via the surrounding gas phase appeared more promising.

Ethanol Atmosphere Storage of cv. *Red Delicious*. When Red Delicious apples were stored in gas-tight jars, the CO_2 concentration increased within 4 days to 3.2 % v/v under the above described conditions, and there was also a slight increase in ester concentrations (Figure 1).
This is in good agreement with other data on the average CO_2 production by stored apples (*18*). In the presence of exogenous ethanol the accumulation of CO_2 was lower (ca. 2 % v/v after 4 days), but the concentrations of all the ethyl esters in the fruit increased dramatically (Figure 1). The most pronounced enrichment was observed for ethyl 3-HO-octanoate: a more than 260fold increase within a period of 48 hours as compared to the content in the untreated control. Apples of the same batch, when stored at ambient atmosphere, lost different proportions of the internal esters and accumulated aliphatic alcohols.
The PA induced ester accumulation, though enzyme catalysed, was not significantly accelerated by elevated storage temperatures (Figure 2a/b). While the longer chain fatty acid esters showed slight increases at 28°C, the concentrations of most volatile esters were decreased as compared to a control sample stored at ambient tempera-

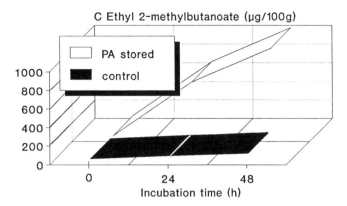

Figure 1. Precursor atmosphere (PA)-storage and concentration of ethyl 2-methylbutanoate (130 mM ethanol/kg apple cv. *Red Delicious*).

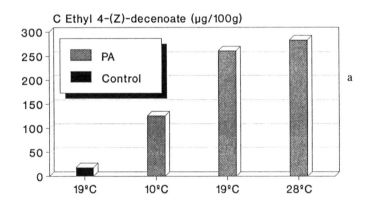

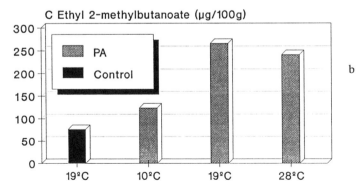

Figure 2a/b. Effect of temperature and PA-storage on concentrations of ethyl esters in apple (100 mM ethanol/kg cv. *Red Delicious*).

ture (19°C). At 10°C, which is closer to the conditions of classical controlled atmosphere storage, all ethyl esters increased as compared to an untreated control sample kept at 19°C.

Comparing different ethanol concentrations in the gas phase, saturation of the esterifying system was reached for short chain fatty acids at concentrations < 50 mM EtOH/kg fruit. At higher ethanol concentrations only the acyl moieties > C5 were affected. Plotting the total neo-synthesis (as a measure of reaction velocity) of ethyl esters vs. ethanol concentration a typical Michaelis curve was obtained reflecting the kinetics of the rate limiting enzyme (Figure 3). There is a small non-linearity of pEtOH over the H_2O/EtOH solution in the applied concentration range that was ignored.

While formally not permissible, the transformed kinetic data (Figure 4) were used to calculate v_{max} and K_M values to compare with known values of pure esterases or lipases. One can conclude that the acylation of the exogenous substrate runs very rapidly, and that ethanol concentrations < 20 mM/kg were quite effective under these conditions. Aged tissues of strawberry converted exogenous l-alanine even somewhat more rapidly with K_M's ranging from 0.26 to 1 mM/kg (19).

In the presence of a sufficient amount of ethanol the increase of the concentrations of the ethyl esters continued for one week or more (Figure 5). The subsequent drop in the concentration of ethyl esters is caused by exhaustion of the precursor, and not by an inactivation of the enzymes involved. Physiological disorders were not observed at this point of time, neither visually nor analytically.

PA-Storage of Apples cv. *Jonathan* **and** *Purple Cousinot*. Butyl and hexyl esters are predominant in these cultivars. As was observed for the incubation of cv. *Red Delicious* with ethanol, the corresponding esters increased during PA-storage at increasing butanol or hexanol concentrations, and again the low molecular weight esters were more affected at lower precursor concentrations (Figure 6a/b). Regarding the total increase of esters the esterifying system is saturated at lower concentrations of precursor with increasing molecular weight of the alcohol. Surprisingly, exogenous alcohol concentrations > 10 mM/kg result in decreases of both internal alcohol and esters contents. According to (20) it appears that the permeability of the peel is a function of the external alcohol concentration. This may be explained as a protective mechanism against stressful external conditions. The prefered formation of the symmetric butyl butanoate and hexyl hexanoate, respectively, points to a concurrent oxidation of the precursor alcohol; the intermediate aldehydes, however, were not detected.

These findings permit the generation of apples with tailored aroma and novel odor notes. A problem may arise,

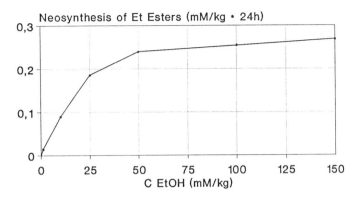

Figure 3. Increase of ethyl esters during PA-storage as f([ethanol]); (cv. *Red Delicious*).

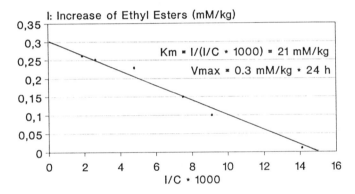

Figure 4. Kinetics of ester neosynthesis in PA-stored apple cv. *Red Delicious*.

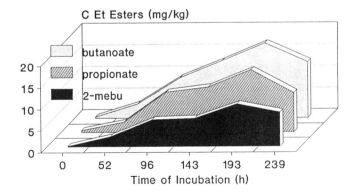

Figure 5. Concentrations of ethyl esters upon long-time PA-storage (50 mM ethanol/kg cv. *Red Delicious*).

if the residual amounts of the precursor alcohol adverse-
ly affect the sensory quality. Experiments with limited
PA-storage periods, followed by a post-incubation at ambient conditions, showed that the internal ratios of precursor alcohol and ester products may be controlled to
ultimately achieve well-balanced sensory results (Figure
7). Due to the continuing consumption of the alcohol, an
immediate drop in the concentration of butanol was observed upon changing the precursor environment (after 24
hours), but the neosynthesis of esters proceeded with a
decline beginning after 3 days of storage. Thus, making
use of the dynamic biosynthetic events, a concerted dilution of the precursor becomes possible.
In fruits containing several sensory key compounds two or
more precursors will have to be used (Figure 8). Esterification of exogenous substrates in fruit appears to be a
highly competitive reaction, as an internal excess of one
alkyl moiety always resulted in a decrease of esters not
containing the respective alkyl moiety. Feeding *Jonathan*
apples with butanol or hexanol clearly showed that the
increase of the respective esters was at the expense of
other volatiles. The resulting sensory deviations from
the genuine aroma were compensated by simultaneously
feeding various alcohols. When compared to an untreated
control, the original molar ratio of esters was restored,
but on a markedly higher level. These ratios were con-
served for several days during prolonged storage at ambient conditions. It is interesting to note that the total molar amount of newly formed esters in the three PA-
experiments of Figure 8 was similar (250 to 270 $\mu M/kg$).
This result underscores the metabolic situation of a
limited pool of endogenous acyl moieties. More refined
incubation protocols considering pROHs, pO_2, and temperature will lead to desired patterns of aroma compounds.

PA-Effects on Primary Metabolites. The acyl moieties required for ester formation originate from a primary pathway. Therefore, changes of concentrations of other,
sensorially or physiologically important constituents
were followed during PA-storage. The concentrations of
sucrose, fructose, and glucose were not markedly
affected. The concentration of total acid went through a
minimum on day 3, but then increased to reach the
original concentration of ca. 10 g/kg after one week.
This time course of acid concentration was similar in PA-
stored and untreated fruit, and depended largely on the
access of oxygen to the fruit. Similarly, the concentration of ascorbic acid dropped from 100 to 55 mg/kg on the
second day of incubation, and then rose continuously, but
again the presence of an alcohol substrate in the gas
phase had little effect on the absolute concentrations.

Recovery of Volatiles from Apple PA. According to Romani
(*21*) the respiratory increase of climacteric fruit is a
measure of the homeostatic reaction of a senescing

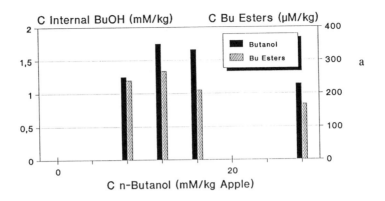

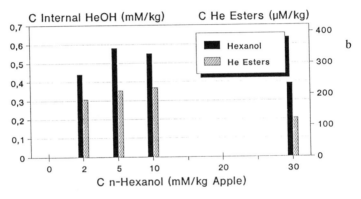

Figure 6a/b. Internal concentration of precursor and total product as f([external precursor]); (cv. *Jonathan*, 3d PA-stored, 23° C).

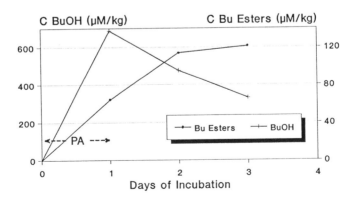

Figure 7. Limited PA-storage and internal concentrations of volatiles (5 mM butanol/kg apple cv. *Cousinot*, 23° C).

tissue. The formation of volatile flavors can be integrated into this view, because the excessive degradation
of membrane polymers and storage compounds implies the
need to eliminate osmotically active and cytotoxic (free
fatty acids) constituents. One means to metabolic relief
is volatilisation and removal from the cell using the
natural concentration gradient. As a result, the headspace of PA-stored fruits contains large amounts of
volatile esters. A fluidised bed filled with a polymer
adsorbent was developed to isolate volatiles from
precursor atmospheres on a lab scale. In one trial 15
consecutive isolations at intervals of 32 hours were
performed (Figure 9). During the first week the concentrations of PA-ennriched esters steadily increased,
followed by a slow decrease during the following two
weeks. The loaded polymer was solvent desorbed, and the
flavor concentrates were sensorically evaluated using
test strips. Even towards the end of the incubation
period typical sensory results with fruity, apple-like
notes were found. A second application of precursor
alcohol was not undertaken, but is expected to maintain
high ester concentrations in the gas phase.

Juices from PA-stored Apples. Aroma enriched fruits
should represent suitable starting materials for
processing. A pilot scale plant for producing juice from
PA-stored and untreated apple was used to demonstrate the
transfer of volatile key compounds from the fruit into a
customary product. During these experiments it turned out
that an adapted technology has to be used: if an immediate inactivation of the endogenous ester hydrolysing
activities is not achieved, severe losses of sensory
valuable compounds will occur. Provided that this technical requirement is met a preceding PA-storage yields juices that contain elevated concentrations of those esters
that determine the fruity character of the over-all
flavor. Table II compares odor thresholds and actual con-

Table II. Volatiles of Juice of PA-treated Apple cv.
Rhein. Bohnapfel (5mM BuOH, 2mM HeOH, 1+2d Storage, 23°C)

Compound (μg/100g)	Threshold[*]	Commercial (average of 3)	PA-stored
n-Butanol	200	1,6	87
n Hexanol	200	3	41
n-Hexanal	2	17	14
2-(E)-Hexenal	32	4	< 1
Butyl acetate	8,8	7	54
Ethyl butanoate	2,7	< 1	41
Hexyl acetate	0,9	15	121

[*] (μg/100g H_2O), Maximum Value acc. to Literature

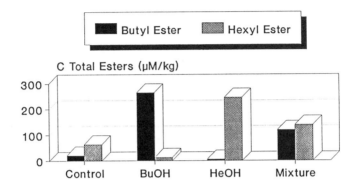

Figure 8. Effect of butanol (5 mM/kg), hexanol (2 mM/ kg), and of a mixture on ester concentrations (1 + 2d PA, apple cv. *Jonathan*, 24° C).

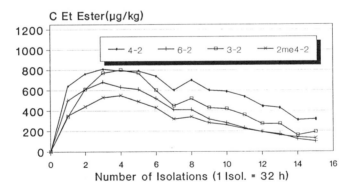

Figure 9. Ethyl esters isolated by adsorption/solvent desorption from apple PA (50 mM ethanol/kg, cv. *Red Delicious*).

centrations of key volatiles in commercial apple juice
and juice made from PA-stored apple. Both precursor alcohols show significantly increased residual concentrations, but remain below the odor detection threshold.
While n-hexanal occurs in all juices in concentrations
above its threshold, 2-(E)-hexenal remains below. Butyl
acetate and ethyl butanoate were found in concentrations
above their thresholds in PA-juices only. The increased
level of hexyl acetate, finally, has a significant impact
on the fruity odor character.
To supplement the analytical data juices from untreated
and PA-stored apples were sensorially evaluated. Commercial juices of good quality served as a reference. At
least 6 out of 9 test persons were able to distinguish
the PA-juices in each of the triangle tests. Most persons
found that the PA-juices were more "apple-like" than the
commercial products. About one third of the test persons
rejected the PA-juice, because, without knowing the
origins of the samples, they argued the unusually intense
odor impressions were caused by artificial flavorings.
In summary, the beneficial effects of PA-storage of apple
were analytically and sensorically demonstrated. The
approach of PA-storage can now be transfered to other
processed products from apple, and, probably, to other
fruits and their products as well.

Acknowledgments. The DECHEMA, Frankfurt, is thanked for
financial support via the Minister of Research and Technology, Bonn. H.Kollmannsberger was involved in some
early PA-research and contributed experimental advice and
GC-MS analyses. Further analytical and technical support
came from A.Keller, G.Leupold, S.Nitz, and I.Reischle.

Literature Cited

1.   Schreier, P. *Chromatographic Studies of Biogenesis
      of Plant Volatiles;* Chromatographic Methods;
      Hüthig: Heidelberg, 1984.
2.   Berger, R.G. In Volatile Compounds in Foods and
      Beverages; Maarse, H., Ed.; Food Science and Tech-
      nology no.44; Dekker: New York, 1991, pp 283-304.
3.   Berger, R.G.; Dettweiler, G.R.; Kollmannsberger,
      H.; Drawert, F. *Phytochemistry* **1990,** *29,* 2069-
      2073.
4.   De Pooter, H.L.; Dirinck, P.J.; Willaert, G.A.;
      Schamp, N.M. *Phytochemistry* **1981,** *20,* 2135-2138.
5.   De Pooter, H.L.; Montens, J.P.; Willaert, G.A.;
      Dirinck, J.P.; Schamp, N.M. *J.Agric.Food Chem.*
      1983, *31,* 813-818.
6.   Berger, R.G.; Drawert, F. *J.Sci.Food Agric.* **1984,**
      35, 1318-1325.
7.   Bartley, I.M.; Stoker, P.G.; Martin, A.D.E.; Hat-
      field, S.G.S.; Knee, M. *J.Sci.Food Agric.* **1985,**
      36, 567-574.

8.   De Pooter, H.L.; van Acker, M.R.; Schamp, N.M. *Phytochemistry* **1987**, *26*, 89-92.

9.   Drawert, F.; Berger, R.G. In *Lebensmittelqualität*; Stute, R., Ed.; VCh: Weinheim, 1988; pp 431-455.

10. Berger, R.G.; Drawert, F.; Kollmannsberger, H. *Z.Lebensm.Unters.Forsch.* **1986**, *183*, 169-171.

11. Guadagni, D.G.; Bomben, J.L.; Harris, J.G. *J.Sci. Food Agric.* **1971**, *22*, 110-115.

12. Willaert, G.A.; Dirinck, P.J.; De Pooter, H.L.; Schamp, N.M. *J.Agric.Food Chem.* **1983**, *31*, 809-813.

13. Dettweiler, G.R. *Erzeugung natürlicher Aromen aus geeigneten Vorstufen mit Hilfe von intakten Fruchtgeweben*; PhD thesis: TU München, 1989.

14. Tressl, R.; Albrecht, W. In *Biogeneration of Aromas*; Parliment, T.H., Croteau, R., Eds.; ACS Symposium Ser. 317; ACS: Washington, DC, 1986; pp 114-133.

15. Knee, M; Hatfield, S.G.S. *J.Sci.Food Agric.* **1981**, *32*, 593-600.

16. Berger, R.G. *Perf.& Flavorist* **1990**, *15*, 33-39.

17. Paillard, N. In *Flavour '81*; Schreier, P., Ed.; deGruyter: Berlin, 1981; pp 479-494.

18. Fidler, J.C.; North, C.J. *J.Hort.Sci.* **1971**, *46*, 213-221.

19. Drawert, F.; Berger, R.G. *Z.Naturforsch.* **1982**, *37c*, 849-858.19.

20. Banks, N.H. *J.Exp.Bot.* **1985**, *36*, 1842-1850.

21. Romani, R.J. *J.Food Biochem.* **1978**, *2*, 221-232.

RECEIVED December 18, 1991

GLYCOSIDIC PRECURSORS

Chapter 7

Glycosidic Precursors of Varietal Grape and Wine Flavor

Patrick J. Williams, Mark A. Sefton, and I. Leigh Francis

The Australian Wine Research Institute, P.O. Box 197, Glen Osmond, South Australia 5064, Australia

The chemical composition and possible biogenetic relationships amongst the norisoprenoid volatiles from hydrolysates of Chardonnay grape precursor fractions are presented along with the sensory contribution that hydrolyzed precursors of Chardonnay and Semillon can make to wines of those varieties. Quantitative descriptive methods were used for the sensory analyses and these showed that attributes such as grassy, tea, lime and honey can be derived from the precursors. The norisoprenoid volatiles found in the precursor hydrolysates can be rationalized as coming from lutein, antheraxanthin, violaxanthin and neoxanthin via 3-hydroxy-α-ionone, 3-hydroxy-β-ionone, 5,6-epoxy-3-hydroxymegastigm-7-en-9-one, and 3,5-dihydroxymegastigma-6,7-dien-9-one.

The interest in flavor development and release in foods has stimulated a growing research activity in flavor precursors (1). In fruits the presence of precursors that could act as a source of latent or potential flavor was recognized many years ago (2,3). These precursors are now known to be non-volatile conjugates of predominantly mevalonic acid- and shikimic acid-derived secondary metabolites. The conjugation found most commonly in fruit flavor precursors is glycosidic, involving glucopyranosides and a range of disaccharide glycosides (4). In most cases these glycosides accumulate in the fruit and can be found at greater concentration than the free aroma constituents.

The major horticultural crop of the world, the grape (*Vitis vinifera*), was among the earliest of many fruits to have been studied for flavor precursors. Accordingly, grapes and wines have possibly benefitted most to date from flavor precursor analysis.

Some grapes have a readily perceived and highly distinctive flavor. These are fruits of the floral varieties, e.g., the various Muscats, Riesling and Gewürztraminer. Free monoterpene compounds are responsible for the floral sensory properties of these grapes, and monoterpene glycosides were first isolated and identified as flavor precursors, both in the berries and in wines made from these varieties (5).

In contrast to the monoterpene-dependent grapes there exist a large number of non-floral varieties that give juices with quite subtle and often non-characteristic aromas. In many cases wines made from these varieties develop distinctive sensory characters only after prolonged maturation. Because of the very low concentration of free aroma

0097–6156/92/0490–0074$06.00/0
© 1992 American Chemical Society

compounds in juices of these non-floral varieties, flavor precursor analysis has been used as a strategy to investigate the flavor compounds (6).

The array of volatile aglycons that has been obtained, by hydrolysis, from glycosides of particular non-floral white wine varieties, e.g., Chardonnay (Sefton, M.A.; Francis, I.L.; Williams, P.J.,The Australian Wine Research Institute, submitted for publication), presents an opportunity to relate, at least in principle, the volatiles and their possible biosynthetic precursors. These relationships are described in the present work. Additionally, the work offers sensory validation of the flavor precursor analysis strategy applied to non-floral white wine varieties. This validation has been achieved by determining the sensory contribution that the total glycosidic precursors, on hydrolysis, are capable of making to wines. In accepting the concept of precursors acting as agents of latent flavor, there is a tacit understanding that hydrolysis of precursors is a process taking place naturally in wines during ageing.

Sensory Descriptive Analysis of the Aroma Properties of Precursor Hydrolysates From Non-Floral Grape Varieties.

Duo-trio difference tests have demonstrated that acid hydrolysates of precursors were readily detected by aroma in a neutral base wine medium (6), and accordingly acid hydrolysates were taken for the current descriptive analysis studies.

For these descriptive analyses, judges were trained initially to quantitatively rate the intensity of eleven aroma attributes using reference standards. Following this, samples of the acid hydrolyzed precursors, dissolved in a neutral base white wine (made from Thompson Seedless grapes) were quantitatively assessed for the same eleven aroma attributes by the judging panel (Francis, I.L.; Sefton, M.A.; Williams, P.J. The Australian Wine Research Institute, submitted for publication). Wines made from the juices from which the precursors were isolated were also evaluated in the study. The results of the descriptive analyses are presented in graphical form in Figures 1 and 2. The center of each graph represents zero intensity, and the distance from the center to the rating point represents the mean of the relative intensity of each sample for that particular attribute. Connecting the mean rating points for a sample allows the profile of the sample to be depicted.

The Sensory Properties of Semillon Precursors.
From the data in Figure 1 it can be seen that the Semillon hydrolysate was significantly enhanced above the base wine in all attributes except estery, toasty and asparagus. Furthermore, with the exception of grassy and asparagus, the Semillon hydrolysate attributes were perceived as significantly greater than, or not significantly different from, those recorded by the panel for the authentic Semillon wine. It is evident that of the five attributes for which this young Semillon wine was perceived as being more intense than the base wine, and which therefore differentiate and characterize it, i.e., toasty, asparagus, grassy, tea and pineapple, the hydrolysis volatiles could be expected to contribute to the last three. Additionally, the precursor hydrolysate attributes which were scored above those of the young wine, i.e., floral, talc, lime, oak, honey and tea, could all be expected to be aromas that develop in this Semillon wine as it ages.

The Sensory Properties of Chardonnay Precursors.
The profile assigned by the panel to the Chardonnay wine (Figure 2) showed that, in relation to the base Thompson Seedless wine, it was significantly more intense in seven of the eleven attributes. Of these seven, tea, lime and honey were three of the four attributes, i.e., floral, tea, lime and honey, that the panel assigned as distinguishing the Chardonnay hydrolysate from the base wine. This suggests that the glycosidic fraction has contributed to these three characterizing attributes of the Chardonnay wine.

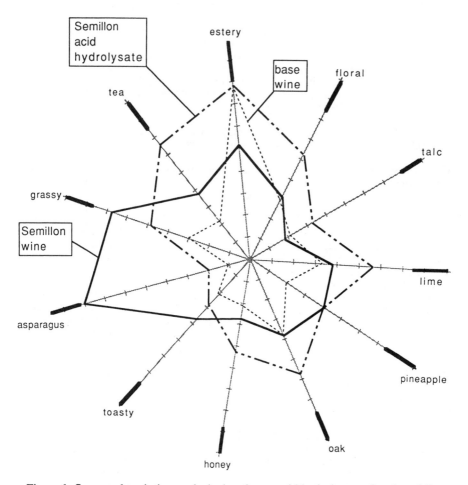

Figure 1. Sensory descriptive analysis data for an acid hydrolysate of a glycosidic precursor fraction from a Semillon juice (assessed in a neutral base wine), a Semillon wine made from the juice, and the neutral base wine. Mean intensity ratings for the eleven attributes were obtained from n=12 judges x 3 replicates for each sample. Least significant differences are represented by heavy bars at the terminus of the attribute axes.

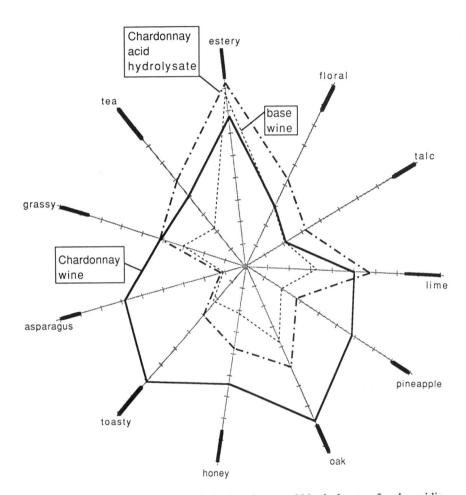

Figure 2. Sensory descriptive analysis data for an acid hydrolysate of a glycosidic precursor fraction from a Chardonnay juice (assessed in a neutral base wine), a Chardonnay wine made from the juice, and the neutral base wine. Mean intensity ratings for the eleven attributes were obtained from n=12 judges x 3 replicates for each sample. Least significant differences are represented by heavy bars at the terminus of the attribute axes.

The Norisoprenoid Aglycons of Chardonnay

Having established that hydrolyzed precursors from juices of these two non-floral grapes have aroma properties with characteristics of the wines prepared from the fruit, the next phase of the research was undertaken. This involved detailed compositional studies of the volatiles obtained from the hydrolyzed grape precursors.

A particular focus of this compositional study was on Chardonnay, and fruit of this variety harvested from the same vineyard over three vintages was used in the research. In excess of three hundred volatile components have been observed to be liberated hydrolytically from the precursor fraction of this grape. Structures have been assigned to more than two hundred of these compounds. The Chardonnay hydrolysates gave volatiles that were particularly rich in norisoprenoid compounds. Many of these have never before been reported as grape constituents, and some are potentially important flavor compounds. The research indicates that, just as floral grape varieties are monoterpene-dependent for their their flavor, Chardonnay can be analogously categorized as norisoprenoid-dependent (Sefton, M.A.; Francis, I.L.; Williams, P.J. The Australian Wine Research Institute, submitted for publication).

The norisoprenoid composition of Chardonnay juices is given in Figures 3 to 6. The volatile norisoprenoids in the Figures are assigned as being glycosidase-released aglycons and/or as aglycons given by acid hydrolysis or as products found free in the juice. The compounds are set out to show their likely biogenetic derivation from known grape carotenoids. Plant carotenoid pigments have been proposed as progenitors of norisoprenoid flavor and aroma compounds in a number of products (7) including tobacco (8) and roses (9).

The Biogenesis of Grape Norisoprenoids From Carotenoids

The great majority of the norisoprenoid compounds found in Chardonnay grape precursor hydrolysates could be derived from the four major xanthophylls reported in grapes, i.e., lutein, antheraxanthin, violaxanthin and neoxanthin (10). It has been reported that grape carotenoid concentration varies with variety (11) and with berry maturity (12). Significantly, carotenoid levels in grape berries were observed to decrease progressively from the onset of fruit development, with the sharpest decline at véraison (12). Whilst little is known of the fate of grape carotenoids on berry ripening, it has been suggested that products of the degradation may be hydrophilic and pass into the juice (12). These hypotheses form the basis of   the biogenetic schemes in Figures 3 to 6. The biogenetic schemes presented in this way also permit a comprehensive compilation of the Chardonnay norisoprenoid constituents to be shown.

The schemes proposed in Figures 3 to 5 involve enzymically catalyzed oxidative cleavage of the 9,10 and 9',10' double bonds of the parent carotenoid (7-9). This leads to four ketones as primary metabolites i.e., 1, 7, 16 and 39, and of these only 3-hydroxy-α-ionone (1) has not been found as a grape glycoside hydrolysis product. The rest of the products that have been observed are derivable from these four ketones by a series of oxidations, dehydrogenations, reductions or eliminations. A few intermediates are inferred in the schemes, but more than 70% of the compounds proposed have been identified as volatiles in the Chardonnay grape glycosidic precursor hydrolysates.

Products From Lutein. (see Figure 3) These can be understood as coming via the two ketones 1 and 7; the latter, 3-hydroxy-β-ionone (7), was identified in both the acid and enzyme-released volatiles. From these two ketones nine other Chardonnay products can be derived, including three dominant volatiles of these hydrolysates, viz, 3-oxo-α-ionol (3), its 4,5-dihydro derivative 5 and the two geometric isomers of 9-hydroxymegastigma-4,6-dien-3-one (13). The last of these are the logical precursors of two products of known aroma significance to tobacco, i.e., megastigmatrienones 14,

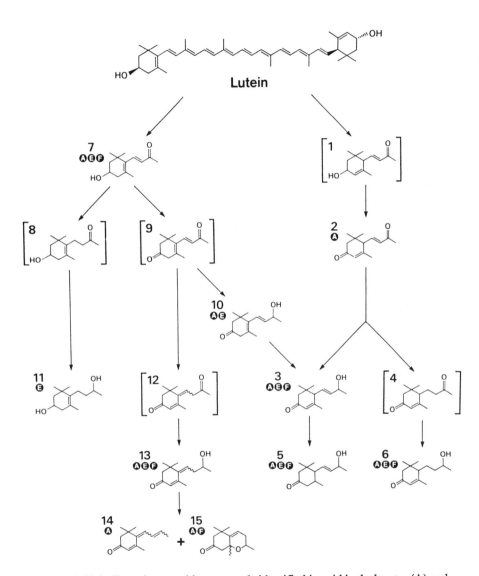

Figure 3. Volatile norisoprenoid compounds identified in acid hydrolysates (A) and glycosidase enzyme hydrolysates (E) of a glycosidic precursor fraction from Chardonnay grapes and also occurring free (F) in the grape juice. Compounds are shown in a hypothetical biogenesis from lutein with inferred intermediates given in square brackets.

which have been described as "the heart of the tobacco aroma" (8) and the oxoedulans 15 which have an oriental tobacco flavor (13).

Products From Violaxanthin. This carotenoid can be seen as providing the genesis of the largest number of Chardonnay norisoprenoids (see Figures 4a and 4b). Thus, reduction of the side chain carbonyl group of the key ketone 16 directs an opening the epoxide ring with participation of the side chain double bond and leads to the 3-oxygenated dihydroactinidol series of compounds 18 and 19. Facile oxidative cleavage of the side chain of these molecules would produce the C_{11} lactones 20 and 21 (see Figure 4a). Alternatively, oxidation of the 3-hydroxyl group of 16 to give the hypothetical intermediate 22 promotes epoxide opening in the other mode to lead, via dehydrovomifoliol (23), to a series of products hydroxylated at position 6 (see Figure 4b). Amongst these are another pair of dominant Chardonnay norisoprenoids, vomifoliol (24) and 4,5-dihydrovomifoliol (25). Reduction of the 3-ketone function of vomifoliol (24) leads, by known pathways (14), to the actinidols (27), the non-megastigmane ketone (28) and 1,1,6-trimethyl-1,2-dihydronaphthalene (TDN) (29).

Reduction of the side chain double bond in the key 6-hydroxylated intermediate, dehydrovomifoliol (23), permits formation by established pathways and via the intermediates 31 and 32, of the known grape and wine volatiles TDN (29) vitispirane (36), and Riesling Acetal (37) (14,15) (Winterhalter, P., The University of Würzburg FRG, personal communication, 1991). While the intermediate polyols 26, 31 and 34 were not observed in this study, these compounds have been found (in rearranged form for 26 and 34) in other grape varieties in which the end products 27-29 and 36-38 are more dominant (14,16) (Winterhalter, P., The University of Würzburg FRG, personal communication, 1991).

Products From Neoxanthin. A simplified form of this pathway given in Figure 5 has been proposed and discussed previously (17). Model hydrolytic studies on a synthetic sample of allenetriol 40, enynediol 41 and the 9-glucoside of 41 (18), as well as on the naturally occurring glycoside of 40 (19), lend strong support to the scheme.

Norisoprenoids From Other Carotenes. Only two norisoprenoids identified in the Chardonnay precursor hydrolysates cannot be accounted for by the pathways from the xanthophylls. These volatiles are shown in Figure 6. The 4-oxo compound 44 and dihydroactinidiolide (45) can both be considered as arising from β-carotene, via, the as yet unobserved, β-ionone, through allylic oxidation for 44 or singlet oxygen-type oxidation of the diene system for dihydroactinidiolide(20). The C_9 ketone 46, which was also identified in the hydrolysates, could come by singlet oxygen-type degradation of C_{13} norisoprenoids derived from the xanthophylls. Several other C_9 ketones observed (but not shown in Figures 3-6) appear to come from acid-catalyzed degradation of higher norisoprenoids in the course of the experiments (21).

There are three points that follow acceptance of the hypothesis that carotenoids are the source of the norisoprenoid volatiles found in these studies.

1) The abundance of thirteen carbon compounds among the grape norisoprenoids indicates that site-specific enzymic catabolism of the carotenoids is the source of the products, rather than non-site-specific chemical type oxidative degradation. The latter mechanism would be expected to give molecules with carbon skeletons containing greater or fewer carbon atoms than the predominantly C_{13} products found.

2) The structural complexity of the volatile products, formed as a result of sequential oxidations and reductions, prove that most of these grape norisoprenoids are genuine metabolites of the fruit. If the various products could have been interrelated by simple

dehydrations and acid-catalyzed rearrangements only, then their genesis may have been purely chemical.

3) The majority of the products shown in Figures 3-6 accumulated in the fruit as glycosides. Presumably the volatiles were produced and transformed before glycosylation, which is consistent with the view that glycosylation is a terminal step in any biosynthetic pathway (*22*). It suggests, furthermore, that glycosides may be transport derivatives for the C_{13} metabolites, allowing them to pass from a lipophilic carotenoid-rich environment into the vacuola sap.

The major norisoprenoids found in the Chardonnay hydrolysates were 3-oxo-α-ionol (3), its 4,5-dihydro derivative 5, the two geometric isomers of 9-hydroxymegastigma-4,6-dien-3-one (13), vomifoliol (24) and 4,5-dihydrovomifoliol (25). It is noteworthy that the first two and last two compounds within this group are related pairs, with the 4,5-double bond being removed to convert the parents, i.e., 3 and 24, to products, i.e., 5 and 25 respectively. Vomifoliol (24) and 3-oxo-α-ionol (3) have been reported as ubiquitous norisoprenoids of *Vitis vinifera* (*21*), and their presence as abundant compounds of these Chardonnay samples was therefore of little surprise. However, the co-accumulation of both of the 4,5-dihydro derivatives of these two ketones, i.e., 5 and 25 in comparable abundance to that of the parents, may be a feature of this variety. In particular, the occurrence in Chardonnay of relatively high concentrations of 4,5-dihydrovomifoliol (25) and 4,5-dihydro-3-oxo-α-ionol (5), highlights possible significant metabolic differences between Chardonnay and other varieties in their ability to produce norisoprenoids.

Riesling glycosidic precursors on hydrolysis give, in addition to monoterpenes, a high concentration of norisoprenoid volatiles including an abundance of Blumenol C (6), TDN (29), vitispirane (36) and Riesling Acetal (37), (*15-17*). These compounds arise via reduction of the side chain double bond of key intermediates 3-oxo-α-ionone (2) and dehydrovomifoliol (23) in the pathways proposed in Figures 3 and 4b. Whilst these normally abundant Riesling norisoprenoids, i.e., Blumenol C (6), TDN (29), vitispirane (36) and Riesling Acetal (37), were found in the hydrolysates of the Chardonnay precursors, they were not major volatiles given by that non-floral variety. It appears that pathways proceeding from 3-oxo-α-ionol (3) and dehydrovomifoliol (23) and involving reduction of the 4,5-double bond rather than side chain and 3-oxo reduction, may be dominant in Chardonnay.

Conclusion

This work was undertaken to establish a sensory justification for the further study of glycosidic fractions from non-floral grape varieties. The work has demonstrated that key aroma attributes of two non-floral wines, including flavors that may be specific to the particular variety, are aromas contributed by the hydrolyzed grape precursors. Additionally, the sensory study on these acid hydrolysates provides some understanding of the flavor development to be expected on cellaring and ageing of these wines.

Having established this sensory justification the volatile norisoprenoids of Chardonnay were examined in detail and a great abundance and diversity of these compounds has been revealed. Although the number of volatiles found precludes sensory studies on individual norisoprenoids, it is noteworthy that many of these compounds have also been found in leaf products such as tobacco and tea (*23*). The assignment by the sensory panel of a significant tea aroma to the Chardonnay hydrolysate is an indication of the likely contribution of the norisoprenoids to this flavor attribute of wine made from Chardonnay.

Finally, there are only limited data available on the catabolic reactions that lead to the disappearance of carotenoids in plants (*24*), and even less is known about the biosynthesis in plants of volatile norisoprenoids. In the absence of formal biosynthetic data from experiments on isotopically labelled carotenoids, the putative pathways

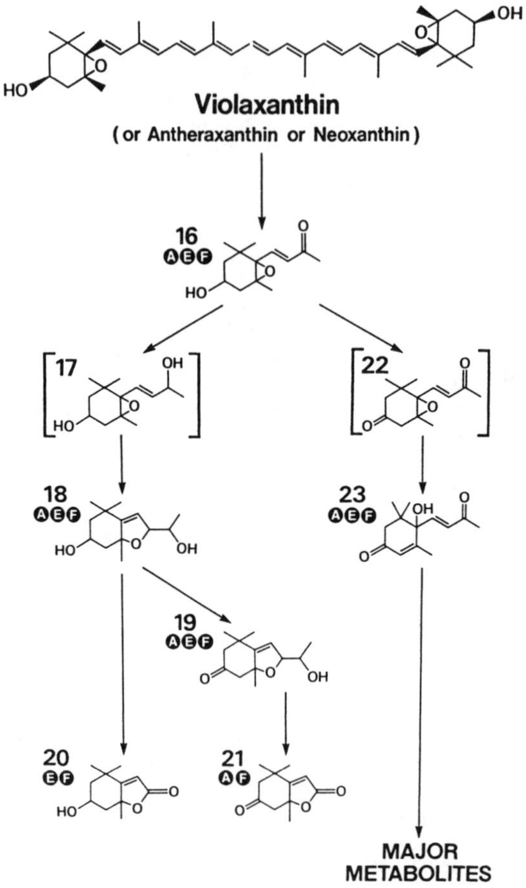

Figure 4a. Volatile norisoprenoid compounds identified in acid hydrolysates (A) and glycosidase enzyme hydrolysates (E) of a glycosidic precursor fraction from Chardonnay grapes and also occurring free (F) in the grape juice. Compounds are shown in a hypothetical biogenesis from violaxanthin, antheraxanthin or neoxanthin, with inferred intermediates given in square brackets.

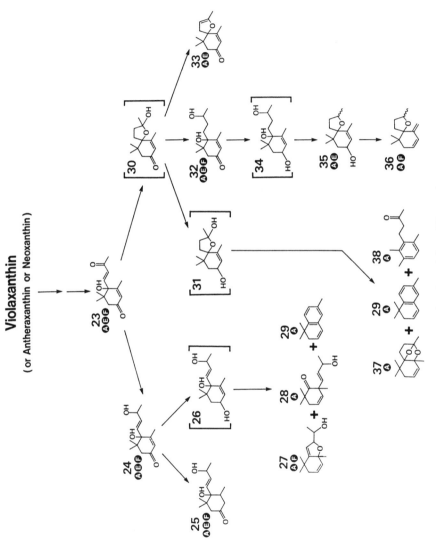

Figure 4b. Continued from Figure 4a.

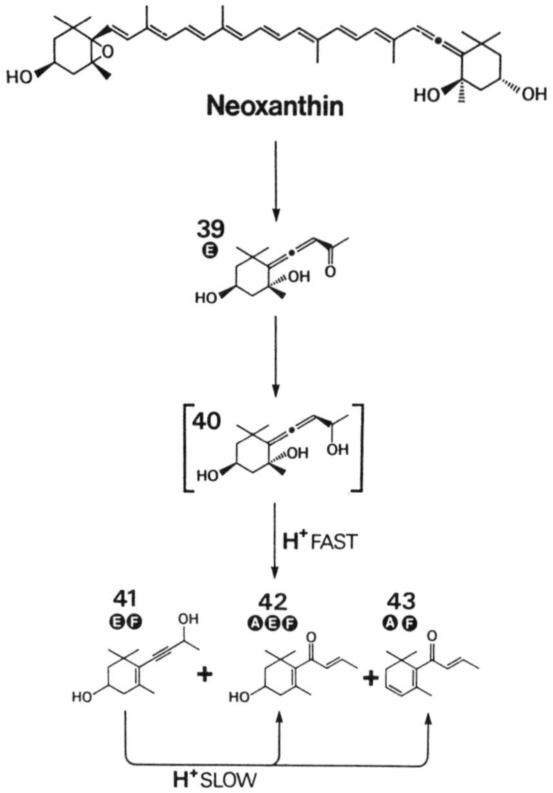

Figure 5. Volatile norisoprenoid compounds identified in acid hydrolysates (A) and glycosidase enzyme hydrolysates (E) of a glycosidic precursor fraction from Chardonnay grapes and also occurring free (F) in the grape juice. Compounds are shown in a hypothetical biogenesis from neoxanthin with inferred intermediates given in square brackets.

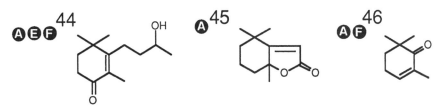

Figure 6.   Some further volatile norisoprenoid compounds identified in acid hydrolysates (A) and glycosidase enzyme hydrolysates (E) of a glycosidic precursor fraction from Chardonnay grapes and also occurring free (F) in the grape juice.

presented here, which are based on consideration of the structures of the many compounds found, offer a working hypothesis to account for the occurrence of norisoprenoids as glycosidic precursors in grapes.

Acknowledgments

We thank G. Gramp & Sons Pty Ltd and The Hardy Wine Company for generously donating samples of grape juices and wines. Staff of the AWRI who participated on the sensory panels are acknowledged and the Grape and Wine Research Council is thanked for supporting the work. Peter Winterhalter of the University of Würzburg FRG, is thanked for valuable discussion and for providing a manuscript of his unpublished work.

Literature Cited

1. Teranishi, R. In *Flavor Chemistry Trends and Developments*; Teranishi, R.; Buttery, R.G.; Shahidi, F., Eds.; ACS Symposium Series No. 388; American Chemical Society: Washington, DC, **1989**; pp. 1-6.
2. Hewitt, E.J.; Mackay, D.A.M.; Konigsbacher, K.; Hasselstrom, T. *Food Technol.* **1956**, *10*, 487.
3. Weurmann, C. *Food Technol.* **1961**, *15*, 531.
4. Williams, P.J. In *Flavor Research Workshop*; Acree T.E., Ed.; ACS Professional Reference Book Series; American Chemical Society: Washington, DC,**1992**, in press.
5. Strauss, C.R.; Wilson, B.; Gooley, P.R.; Williams, P.J. In *Biogeneration of Aromas*; Parliment, T.H.; Croteau, R., Eds.; ACS Symposium Series No. 317; American Chemical Society: Washington, DC, **1986**; pp. 222-242.
6. Williams, P.J.; Sefton, M.A.; Wilson, B. In *Flavor Chemistry Trends and Developments*; Teranishi, R.; Buttery, R.G.; Shahidi, F., Eds.; ACS Symposium Series No. 388; American Chemical Society: Washington, DC, **1989**; pp. 35-48.
7. Weeks, W.W. In *Biogeneration of Aromas*; Parliment, T.H.; Croteau, R., Eds.; ACS Symposium Series No. 317; American Chemical Society: Washington, DC, **1986**; pp. 157-166.
8. Enzell, C. *Pure & Appl. Chem.* **1985**, *57*, 693-700.
9. Eugster, C.H.; Märki-Fischer, E. *Angew. Chem. Int. Ed. Engl.* **1991**, *30*, 654-672.
10. Gross, J. *Pigments in Fruits* ; Academic Press: London, **1987**; pp. 87-186.
11. Razungles, A.; Bayonove, C.L.; Cordonnier, R.E.; Baumes, R.L. *Vitis* **1987**, *26*, 183-191.
12. Razungles, A.; Bayonove, C.L.; Cordonnier, R.E.;Sapis, C.J. *Am. J. Enol. Vitic.* **1988**, *39*, 44-48.

13. Lloyd, R.A.; Miller, C.W.; Roberts, D.L.; Giles, J.A.; Dickerson, J.P.; Nelson, N.H.; Rix, C.E.; Ayers, P.H. *Tob. Sci* . **1976**, *20,* 40-48.
14. Strauss, C.R.; Dimitriadis, E.; Wilson, B.; Williams, P.J. *J. Agric. Food Chem.* **1986**, *34,* 145-149.
15. Winterhalter, P.; Sefton, M.A.; Williams, P.J. *Chem. Ind. (London)* **1990**, 463-464.
16. Winterhalter, P.; Sefton, M.A.; Williams, P.J. *J. Agric. Food Chem.* **1990**, *38,* 1041-1048.
17. Winterhalter, P.; Sefton, M.A.; Williams, P.J. *Am. J. Enol. Vitic.* **1990**, *41,* 277-283.
18. Skouroumounis, G.K. Beta Damascenone Precursors in Grapes and Wines. Ph.D. Thesis, The University of Adelaide, Adelaide, South Australia, **1991**.
19. Näf, R.; Velluz, A.; Thommen, W. *Tetrahedron Lett.* **1990**, *31,* 6521-6522.
20. Wahlberg, I.; Enzell, C.R. *Nat. Prod. Rep.* **1987**, *4,* 237-276.
21. Strauss, C.R.; Wilson, B.; Williams, P.J. *Phytochemistry* **1987**, *26,* 1995-1997.
22. Hösel, W. In *The Biochemistry of Plants. Secondary Plant Products;* Conn, E.E., Ed.; Academic: London, **1981**; Vol. 7, pp. 725-753.
23. Ohloff, G. In *Progress in the Chemistry of Organic Natural Products;* Herz,W.; Grisebach, H.; Kirby, G.W. Eds.; Springer Verlag: New York, **1978**; Vol. 35, pp. 431-527.
24. Britton, G. *Nat. Prod. Rep.* **1989**, *6,* 359-392.

RECEIVED December 18, 1991

Chapter 8

Glucosides of Limonoids

Shin Hasegawa[1], Chi H. Fong[1], Zareb Herman[1], and Masaki Miyake[2]

[1]Fruit and Vegetable Chemistry Laboratory, U.S. Department of Agriculture, 263 South Chester Avenue, Pasadena, CA 91106
[2]Wakayama Agricultural Biological Research Institute, Tsukatsuki, Momoyama, Wakayamaken, Japan 649-61

Limonoids are one of the two bitter principles in *Citrus*. Recently, limonoids have been shown to be present as glucoside derivatives in the Rutaceae. The 17-ß-D-glucopyranosides of 17 citrus limonoids have been isolated from *Citrus* and its hybrids. Four such compounds have been also isolated from non-citrus members of the Rutaceae. They are all nonbitter. The glucosidation of limonoids is an enzymic conversion of bitter to nonbitter compounds and is a natural limonoid debittering process taking place in *Citrus*. This chapter describes the discovery, occurrence, biosynthesis and possible biological significance of limonoid glucosides in *Citrus*.

Limonoids are a group of chemically related triterpene derivatives found in the Rutaceae and Meliaceae families. Among 37 limonoid aglycones reported to occur in *Citrus* and its hybrids, four are known to be bitter in taste. These are limonin, nomilin, ichangin and nomilinic acid (*1*). Limonin is the primary cause of citrus juice bitterness. This problem occurs in a variety of citrus juices, and it is economically important to the citrus industry because bitter juices have a lower market value for producers.

Bitterness occurs generally in juice extracted from fruits harvested in early season, but is greatly reduced later in the season. This is due to a decrease in the concentration of limonoids in fruit tissues as fruit maturation progresses. However, how limonoids are metabolized in fruit during late stages of fruit growth and maturation was not understood until recently when Hasegawa et al.(*2*) discovered that limonoids also occur in *Citrus* as nonbitter glucoside derivatives. The discovery of limonoid glucosides finally showed that limonoid aglycones are converted to their glucosides during late stages of fruit growth and maturation.

At least 17 limonoid glucosides from *Citrus* (*2-8*) and 4 from non-citrus members of the Rutaceae family (*9,10*) have been isolated. Limonoid glucosides are also found to be one of the major secondary metabolites present in *Citrus* (*5,11*) (e.g. 320

0097–6156/92/0490–0087$06.00/0
© 1992 American Chemical Society

ppm in commercial orange juices and 0.6% of dry weight in
seeds), and they are biosynthesized in fruit tissues and seeds
during late stages of fruit growth and maturation (12).
Limonoid UDP-D-glucose transferase catalyzes the glucosidation
of limonoids in *Citrus* and its activity occurs only in matured
fruit tissues and seeds (12-14). The glucosidation of limonoids
is considered to be a natural limonoid debittering process
taking place in *Citrus.*

Delayed Bitterness

Most citrus fruits are not bitter tasting if eaten fresh or if
freshly squeezed juice is consumed. However, a few hours after
juicing,the juice extracted from a variety of citrus fruits
becomes bitter. This phenomenon is known as delayed
bitterness, and the mechanism is shown in Figure 1.

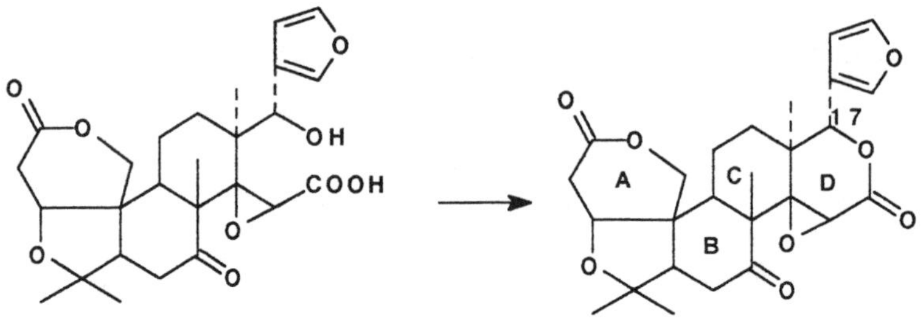

Limonoate A-ring lactone   Limonin
(Nonbitter Precursor)   (Bitter)

Figure 1. Mechanism of Delayed Bitterness

Intact citrus fruits do not contain bitter limonin, but
rather a nonbitter precursor of limonin, limonoate A-ring
lactone (LARL)(15). When the fruit is juiced, LARL is
gradually converted to limonin under acidic conditions. This
conversion is accelerated by the action of an endogenous
enzyme, limonin D-ring lactone hydrolase. The resulting
limonin is extremely bitter in taste. Most persons can detect
limonin bitterness at 5 or 6 ppm in juice. Juice extracted
from early to mid-season navel oranges may contain over 30 ppm
limonin. However, juice from fruit harvested late in the
season is usually nonbitter because it contains less than 6 ppm
of limonin.

Discovery of Limonoid Glucosides

The procedure which has been used by the citrus industry and
research laboratories to analyze or isolate limonoids from
citrus tissues and juices involves extraction with organic
solvents such as methylene chloride and ethyl acetate. The
aqueous fraction has been ignored because limonoids are
extracted in the organic fractions. Recently, we examined the
aqueous extracts of citrus tissues and juices, and we
discovered that *Citrus* contains large quantities of water
soluble limonoid derivatives. (See Figure 2.)
     The first compound was isolated from an extract of
grapefruit seeds by column chromatography and identified as
limonin 17-ß-D-glucopyranoside, known as limonin glucoside

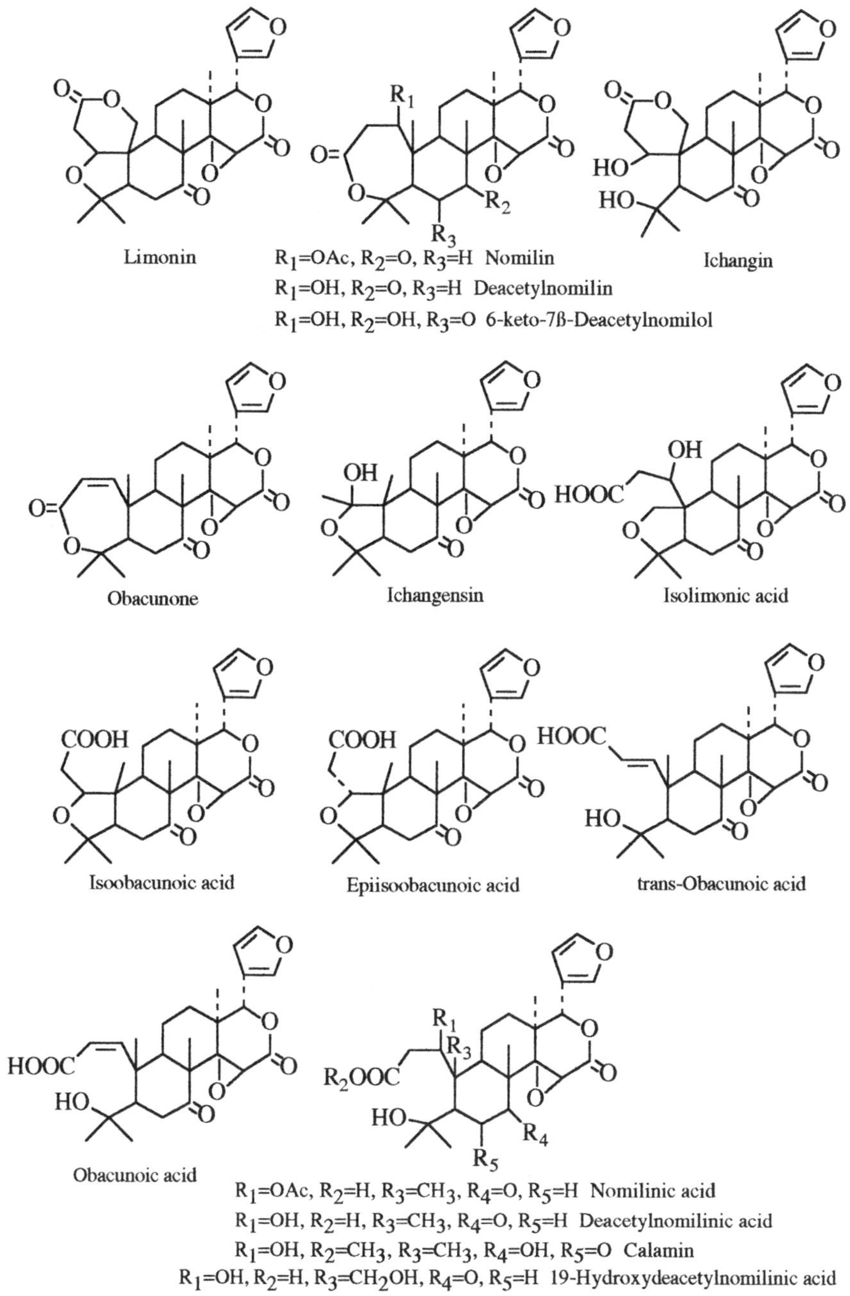

Limonin

R_1=OAc, R_2=O, R_3=H Nomilin
R_1=OH, R_2=O, R_3=H Deacetylnomilin
R_1=OH, R_2=OH, R_3=O 6-keto-7ß-Deacetylnomilol

Ichangin

Obacunone

Ichangensin

Isolimonic acid

Isoobacunoic acid

Epiisoobacunoic acid

trans-Obacunoic acid

Obacunoic acid

R_1=OAc, R_2=H, R_3=CH$_3$, R_4=O, R_5=H Nomilinic acid
R_1=OH, R_2=H, R_3=CH$_3$, R_4=O, R_5=H Deacetylnomilinic acid
R_1=OH, R_2=CH$_3$, R_3=CH$_3$, R_4=OH, R_5=O Calamin
R_1=OH, R_2=H, R_3=CH$_2$OH, R_4=O, R_5=H 19-Hydroxydeacetylnomilinic acid

Figure 2. Structures of Limonoids

(Figure 4) (2). One limonin molecule is linked with one
D-glucose molecule at the 17-position of the open D-ring by a
ß-glucosidic linkage. Using column chromatography and NMR
spectroscopy, seventeen of these compounds have been isolated
and identified from *Citrus* and its hybrids (2-8). These are
listed in Table 1.

Table 1. Limonoid Glucosides in *Citrus*

Monocarboxylic acids	Dicarboxylic acids
17-ß-D-glucopyranosides of:	17-ß-D-glucopyranosides of:
1. limonin	10. nomilinic acid
2. nomilin	11. deacetylnomilinic acid
3. deacetylnomilin	12. obacunoic acid
4. obacunone	13. trans-obacunoic acid
5. ichangin	14. isoobacunoic acid
6. ichangensin	15. epiisoobacunoic acid
7. calamin	16. 19-hydroxydeacetyl-
8. 6-keto-7β-deacetylnomilol	nomilinic acid
9. methyl deacetylnomilinate	17. isolimonic acid

All of the compounds isolated are either monocarboxylic or
dicarboxylic acids. All contain one glucose moiety in the same
position as limonin 17-β-D-glucopyranoside. Since 37 limonoid
aglycones have been identified, it is very possible that each
aglycone has a corresponding glucoside derivative. All
limonoid glucosides isolated to date are nonbitter to taste.
   Limonoid glucosides are also found to be present in non-
citrus members of the Rutaceae family. The 17-ß-D-glucopyrano-
sides of limonin, limonin diosphenol and 6ß-hydroxy-5-
epilimonin are present in *Tetradium rutaecarpa* (9). The 17-ß-
D-glucopyranosides of limonin and obacunone have been isolated
from *Phellodendron amurense* (10).

Analysis of Limonoid Glucosides

TLC is the best method of detecting limonoid glucosides. It can
be used to quantify limonin glucoside and the total limonoids
(11). However, it is difficult to separate individual
glucosides by the method used thus far. The solvent system used
is EtOAc-methyl ethyl ketone-formic acid (88%)-H_2O (5: 3: 1: 1)
(11). The method involves spotting the sample onto a silica
gel plate along with the appropriate limonoid glucoside
standard, developing the plate in a suitable solvent system,
spraying the plate with Ehrlich's reagent (1% p-dimethylamino-
benzaldehyde in EtOH), and developing the plate in a chamber of
HCL gas. Limonoid glucosides produce distinctive reddish-orange
spots on the plate.
   A HPLC method for the quantification of limonin glucoside
in citrus juices was first developed by Fong et al (11).
Later, Herman et al (16) reported a method for the analysis of
individual limonoid glucosides in orange juice. Ozaki et al.

(5) reported a method for analysis of seven limonoid glucosides of citrus seeds. The method uses a C18 reverse-phase analytical column and a linear gradient starting with 15% CH3CN in 3 mM H_3PO_4 and ending with 26% CH_3CN at 33 min. The flow rate was 1 ml per min and the glucosides are detected by UV absorption at 210 nm (5).

Extraction and Isolation of Limonoid Glucosides

Since limonoid glucosides are water soluble, they can be extracted with water from citrus seeds. It is possible to extract both limonoid aglycones and glucosides from the same batch of citrus seed meal (2,3). The meals are washed with hexane or petroleum ether to remove oily materials. Aglycones are extracted with acetone. Glucosides are then extracted with MeOH. The MeOH extract also contains some aglycones which can be separated by partitioning between water and CH_2Cl_2. The water fraction containing glucosides can be used for isolation of individual glucosides.

Limonoid glucosides can be isolated by open column chromatography, preparative HPLC or a combination of both (2-8). Since crude extracts obtained from seeds or fruit tissues contain flavonoid glucosides which co-elute with limonoid glucosides during column chromatography, a DEAE Sephacel column is helpful to separate these two types of glucoside before subsequent chromatographic separation. Flavonoid glucosides have no affinity to the column, whereas limonoid glucosides have weak affinity and can be eluted with 0.2 M NaCl solution. XAD and XAD-2 have been used for open column chromatography, and a C18 reverse-phase column has been used for preparative HPLC chromatography. In both methods, either MeOH or CH_3CN has been used for eluting limonoid glucosides.

Identification of limonoid glucosides

The most useful method for obtaining structural information about limonoid glucosides is nuclear magnetic resonance (NMR) spectroscopy (2,3). These spectra are run in deuterated dimethyl sulfoxide, in which the glucosides are highly soluble, at 90°. At lower temperatures the signals are severely broadened. The ^{1}H NMR spectrum of a new limonoid glucoside, when compared with spectra of known limonoids, can provide a good indication of the possible structure of the aglycone. The ^{13}C NMR spectrum then usually confirms the structure. Typically the signals of carbons near the 17-position are considerably shifted, while those further away are almost identical to those of the aglycone. In difficult cases running 2D NMR spectra may be necessary to determine the structures. In particular, the linkage of the sugar to the 17-position is shown by a cross peak between H-17 and glucose H-1 in the 2D NOESY spectrum. For identification of known limonoid glucosides the ^{1}H NMR spectrum is sufficient. The positions of the four or five C-methyl signals are significantly different for each known glucoside.

Occurrence

Limonoid glucosides are one of the major secondary metabolites in *Citrus*. These compounds are present in fruit tissues and seeds in high concentrations (5,11,16). However, they are not present in leaves, stems, and immature fruit

tissues and seeds (14). Commercial citrus juices contain very
high concentrations of limonoid glucosides (11). Analyses of 15
orange, 8 grapefruit and 4 lemon juices showed that orange
juice contained the highest amounts, averaging 320 ppm of total
limonoid glucosides, ranging from 250 to 430 ppm. Since orange
juice contains generally 2 ppm of limonoid aglycones (1), this
is 125 to 215 fold higher than the aglycones. Grapefruit juice
averaged 190 ppm, ranging from 140 to 230 ppm. Lemon juice
contained the lowest level averaging 82 ppm, ranging 76 to 93
ppm.
      Limonin glucoside is the major limonoid glucoside in
commercial juices (11). Orange juice contains an average of
180 ppm or 56% of the total. Grapefruit juice averaged 54 ppm,
which is 63% of the total, while lemon juice averages 54 ppm
which is 66% of the total. Analyses of individual limonoid
glucosides in orange juice show that limonin glucoside is the
predominant, followed by glucosides of nomilinic acid, nomilin,
deacetylnomilinic acid, deacetylnomilin and obacunone in order
of decreasing concentration (16).
      Citrus seeds also contain very high concentration of
limonoid glucosides. In fact, these compounds comprise nearly
1% of the fresh weight of grapefruit seeds (2). Ozaki et
al.(5) determined the concentrations of individual limonoid
glucosides in the seeds of eight citrus species. As shown in
Table 2, the total concentration of limonoid glucosides in the
seeds ranges from 0.33 to 0.89% and averages 0.61% of the dry
weight. This concentration is approximately 20-fold higher
than in juice.

Table 2. Limonoid Glucosides in Citrus Seeds*

Seeds	DAG	NG	NAG	OG	LG	DG	Total
Fukuhara	0.28	3.22	0.98	1.09	0.51	1.32	7.40
Hyuganatsu	0.42	1.10	0.76	0.65	-	0.37	3.31
Sanbokan	0.37	1.13	0.55	0.90	0.51	0.89	4.36
Shimamikan	0.48	1.89	1.29	2.35	0.37	0.69	7.08
Grapefruit	0.75	2.01	0.89	0.86	1.48	0.68	6.67
Lemon	0.14	1.53	1.39	1.49	1.44	0.55	6.54
Valencia	0.13	4.48	0.98	1.06	0.59	1.69	8.94
Tangerine	1.69	0.42	0.96	0.45	0.90	0.93	5.36

      * Determined by HPLC, mg/g of dry seeds.
      G, glucoside; DA, deacetylnomilinic acid; N, nomilin; O,
obacunone; L, limonin, D, deacetylnomilin (5).

      Unlike fruit tissue, where limonin glucoside predominates
(11, 16), the concentration of limonin glucoside is relatively
low in the seeds (5). Nomilin glucoside is the major glucoside
in the majority of the seeds. Particularly, in Fukuhara and
Valencia orange, nomilin glucoside makes up 42 and 51% of the
total limonoid glucosides, respectively. The glucosides of
obacunone and deacetylnomilin are the major glucosides in
Shimamikan and tangerine, respectively.
      Citrus seeds are excellent sources of both limonoid
aglycones and glucosides. For example, the concentration of
total limonoids in grapefruit seeds is approximately 3% of the
dry weight (2,5,17). The concentration of limonoid glucosides
in the seeds is generally lower than that of limonoid

aglycones. On average, the ratio of total aglycones to total glucosides is approximately 2 to 1.

Citrus fruit peels also contain high concentrations of limonoid glucosides. Their concentrations increase as fruit maturation progresses. Peels obtained from navel oranges grown in California and harvested on 11/2, 11/30/1987, 1/4, 2/6, 3/16 and 4/10/1988 contained 66, 202, 298 ,258, 286 and 362 ppm of limonin glucoside, respectively (12). Valencia orange peels harvested on 11/4/1988, 1/31, 3/10, 4/21, 6/22 and 7/28/1989 contained 186, 396, 426, 433, 523 and 733 ppm of total limonoid glucosides, respectively (10).

Biosynthesis of Limonoid Glucosides

As soon as limonoid glucosides were discovered, it was postulated that they were biosynthesized from their open D-ring limonoid precursors. This hypothesis was confirmed by radioactive tracer experiments in which ^{14}C-labelled nomilin was converted to nomilin 17-ß-D-glucopyranoside in lemon seeds and albedo (14) and in navel orange albedo (18). Other tissues, such as leaves, stems, and immature fruit tissues and seeds exhibited no capacity to biosynthesize limonoid glucosides(14).

The concentration of limonoate A-ring lactone (LARL), which is a precursor of limonin and is the predominant limonoid in citrus fruit tissues, decreases as fruit maturation progresses (Figure 3) (12). This compound has been shown to convert to other minor limonoids such as 17-dehydrolimonoate A-ring lactone (19), deoxylimonin and deoxylimonic acid (20) and possibly to limonol and limonyl acetate. However, these compounds are very minor and alone cannot explain the decrease of LARL during maturation. The discovery of limonin glucoside and the formation of limonin glucoside finally explain how LARL disappears at late stages of fruit growth and maturation.

Figure 3 shows that in navel oranges grown in California the initiation and subsequent increase in limonin glucoside biosynthesis is accompanied by a simultaneous decrease in LARL content (12). During September when the fruits were still green, limonin glucoside began to appear in the flesh portion, and the content increased sharply thereafter. The sudden increase in the limonin glucoside content of navel oranges during September simultaneous with a sudden decrease in LARL confirmed that limonin glucoside is formed by the glucosidation of LARL. Since limonin glucoside is nonbitter, the glucosidation of LARL to form limonin glucoside would be a natural debittering process. The stimulation of this conversion through the use of plant bioregulators or through genetic engineering could reduce the content of LARL and subsequently reduce the limonoid bitterness in citrus juices.

As shown in Figure 3, the total limonin (LARL + limonin glucoside) content increased sharply from June to August, slowed down once in September, increased sharply again during October and November, and remained fairly constant thereafter. This showed unexpectedly that the biosynthesis of limonoids continues to occur in the fruit tissues until the end of November. It is of interest to note that the total weight of the fruit increased also until the end of November. Hasegawa et al. (12) also analyzed total limonoid glucosides and found that ratios of limonin glucoside to total limonoid glucosides at different stages of fruit growth and maturation were fairly

constant, 0.7.   These data suggest that all limonoid aglycones
are simultaneously converted to their glucosides throughout
late stages of fruit growth and maturation.

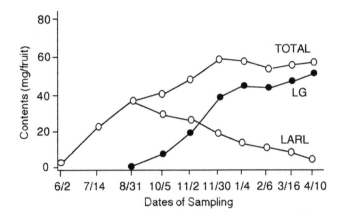

Figure 3. Changes in the limonoate A-ring lactone (LARL) and
limonin 17-ß-D-glucopyranoside (LG) contents of navel oranges
during fruit growth and maturation. LG values are expressed as
LARL by multiplying by 0.723 (12).

        The limonoids present in seeds are most likely
biosynthesized independently from the biosynthesis occurring in
fruit tissues.   The composition of limonoid glucosides in the
seed is quite different from that present in the fruit tissue.
For example, limonin glucoside is the predominant limonoid
glucoside in the fruit tissue (11,16), but in the seed its
concentration is very low (5).   Instead, nomilin glucoside is
the major glucoside in the seed. Seeds possess high
concentrations of both limonoid aglycones and glucosides.   The
ratio of total aglycones to total glucosides in the seed
averages 2 to 1.   In the juice this ratio is about 1 to 150. In
addition to the above differences between seeds and fruit
tissues, there are at least two more major biochemical
differences between them. First, the dilactones are the
predominant aglycones in the seed, whereas the monolactones are
predominant in the fruit tissue. Secondly, during late stages
of fruit growth and maturation, the aglycone content in the
fruit tissue decreases due to its conversion to glucosides
(12).   However, this decrease does not occur in the seeds (10)
since the predominant dilactone form cannot be converted to
glucosides.   These data strongly suggest that limonoid and
their glucoside biosynthesis in seeds and fruit tissues occurs
independently.
        The enzyme responsible for the glucosidation of limonoids
has been identified as limonoid UDP-D-glucose transferase
(Figure 4) (13). The activity has been demonstrated in cell-
free extracts of citrus albedo and seed tissues. When [14]C-
labelled nomilinoate A-ring lactone was incubated with cell-
free extracts of orange albedo, it was converted to labelled

nomilin 17-ß-D-glucopyranoside. The crude enzyme preparations required UDP-D-glucose as substrate(*10*). This enzyme has yet to be isolated.

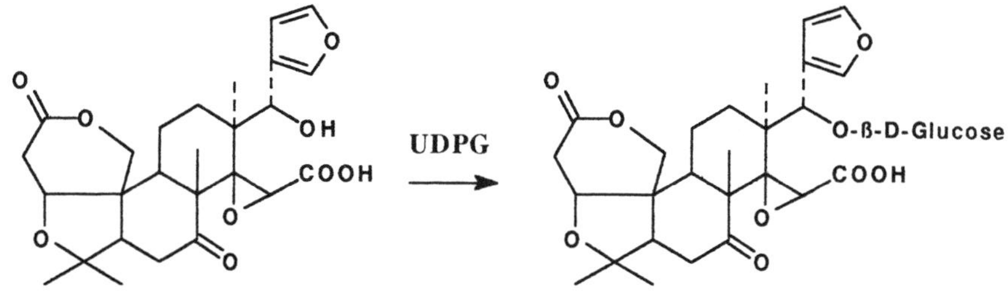

Limonoate A-ring lactone   Limonin 17-ß-D-glucopyranoside
                                                      (Nonbitter)

Figure 4.   Limonoid UDP-D-glucose Transferase

Biodegradation of Limonoid Glucosides

There is no evidence that limonoid glucosides are converted back to aglycones after they are formed in the fruit (*18*). Radioactive tracer work shows that C-labelled nomilin glucoside was not metabolized in albedo of navel oranges and lemons. However, during germination, the limonoid glucosides in seeds appear to be metabolized.

A species of bacterium isolated from soil by enrichment on limonin glucoside as a sole carbon source possesses ß-glucosidase which exhibits activity on limonoid glucosides (*2,8,10*). The enzyme attacks all of the limonoid glucosides isolated from the Rutaceae. No commercially available glucosidase is active on limonoid glucosides.

Chemical Stability of Limonoid Glucosides

Most of the limonoid glucosides appear stable at pH 2 to pH 8, but this has not been well studied. There is no significant difference in the total and individual limonoid glucosides between freshly prepared and concentrated orange juices (*11*). One known exception is nomilin glucoside, which converts to nomilinic acid glucoside below pH 3, and to obacunone glucoside at pH 7.5 and above (*10*).

Biological Activities of Limonoids

Limonoids are also being investigated for possible anti-cancer effects. Certain limonoids have been found to induce glutathione S-transferase, a detoxifying enzyme, in the liver and small intestinal mucosa of mice and to inhibit benzo(a)pyrene-induced neoplasia in forestomach of mice (*21-22*). They also inhibit the formation of 7,12-dimethyl-benz[a]anthracene(DMBA)-induced buccal pouch epidermoid carcinomas in hamsters (*23*).

More recently, limonin glucoside has also been found to inhibit tumor growth in hamsters (*24*). Limonin glucoside exhibited a significant decrease in average tumor burden in hamster buccal pouch induced by DMBA. However, nomilin and nomilinic acid glucosides had no significant effects on the DMBA-induced oral carcinogenesis.

The biological function of limonoids in plants is not yet known. However, there is considerable evidence that limonoids act as antifeedants against a variety of insect pests (25-31). Thus, limonoids may serve as a protection against insect damage.

Conclusion

Significant progress has been made in biochemical research on citrus limonoids. We know now where, when, and how limonoids and their glucoside derivatives are biosynthesized, metabolized and accumulated in *Citrus*. The discovery and formation of limonoid glucosides finally explain how limonoid aglycones, such as bitter limonin, are metabolized and disappear in the fruit during late stages of fruit growth and maturation. The stimulation of this natural debittering process by bioregulation could reduce the content of the precursors of bitter limonoids and subsequently reduce the limonoid bitterness in citrus juices.

Limonoids have been shown to have anticarcinogenic activity in laboratory animals. Limonoid glucosides appear to have the similar activity. However, these investigations are still at preliminary stages, and further research is needed to determine whether levels of limonoid aglycones and glucosides present in citrus fruit have significant effects as anti-cancer dietary sources. On the other hand, citrus seeds contain very high concentrations of both limonoid aglycones and glucosides, and are excellent sources of such compounds.

Acknowledgment

We thank Dr.Raymond Bennett for helpful discussions in the preparation of this manuscript.

Literature Cited

1. Maier, V. P.,; Bennett, R. D.; Hasegawa, S. In *Citrus Science and Technology*. Ed. Nagy, S.; Shaw, P. E.; Veldhuis, M. K. Avi Publishing Co., Westport, CT, 1977, Vol 2. 355-396.
2. Hasegawa, S.; Bennett, R.D.; Herman, Z., Fong, C.H.; Ou, P. *Phytochemistry* 1989, *28*, 1717.
3. Bennett, R.D.; Hasegawa, S.; Herman, Z. *Phytochemistry* 1989, *28*, 2777.
4. Maeda, H.; Ozaki, Y.; Miyake, M.; Ifuku, Y.; Hasegawa, S. *Nippon Nogeikagaku Kaishi* 1990, *64*, 1231.
5. Ozaki, Y.; Fong, C.H.; Herman, Z.; Maeda, M.; Miyake, M.; Ifuku, Y.; Hasegawa, S. *Agric. Biol. Chem.* 1991, *55*, 137.
6. Ozaki, Y.; Miyake, M.; Maeda, H.; Ifuku, Y.; Bennett, R.D.; Herman, Z.; Fong, C.H.a; Hasegawa, S. *Phytochemistry* 1991, *30*, 2659.
7. Bennett, R. D.; Miyake, M.; Ozaki, Y.; Hasegawa, S. *Phytochemistry* 1991 (in press)
8. Miyake, M.; Ozaki, Y.;; Ayano, S.; Bennett, R. D.; Herman, Z.; Hasegawa, S. *Phytochemistry* 1991 (in Press)
9. Ozaki, Y.; Miyake, M.; Maeda, H.; Ifuku, Y.; Bennett, R.D.; Hasegawa, S. *Phytochemistry* 1991, *30*, 2365.

10. Hasegawa, S.; Miyake, M.; Fong, C.H. Unpublished data, Fruit and Vegetable chemistry Laboratory, USDA, ARS, Pasadena CA 91106
11. Fong, C.H.; Hasegawa, S.; Herman, Z.; Ou, P. *J. Food Sci.* **1990**, *54*, 1501.
12. Hasegawa, S.; Ou, P.; Fong, C.H.; Herman, Z.; Coggins, C.W., Jr.; Atkin, D.R. *J. Agric. Food Chem.* **1991**, *39*, 262.
13. Herman, Z.; Fong, C.H.; Hasegawa, S. Coggins, C.W.,Jr.; Atkin, D.R. Abstract, Citrus Research Conference, Pasadena, CA., 1990, P. 8.
14. Fong, C.H.; Hasegawa, S.; Herman, Z.; Ou, P. *J. Sci. Food Agric.* **1991**, *54*, 393.
15. Maier, V.P.; Beverly, G.D. *J. Food Sci.* **1968**, *33*, 488.
16. Herman, Z.; Fong, C.H.; Ou, P.; Hasegawa, S. *J. Agric. Food Chem.* **1990**, *38*, 1860.
17. Hasegawa, S.; Bennett, R. D.; Verdon, C.P. *J. Agric. Food Chem.* **1980**, *28*, 922.
18. Herman, Z.; Fong, C.H.; Hasegawa, S. *Phytochemistry* **1991**, *30*, 1487.
19. Hasegawa,S.; Maier, V. P. and Bennett,R. D. *Phytochemistry* **1980**, *13*, 103.
20. Hasegawa, S.; Bennett, R. D.; Verdon,C. P. *Phytochemistry* **1974**,*19*,1445.
21. Lam, L.K.T.; Hasegawa, S. *Nutrition and Cancer*, **1989**, *12*,43.
22. Lam, L.K.T.; Li, Y.; Hasegawa, S. *J. Agric. Food Chem.,* **1989**, *37*, 878.
23. Miller, E.G.; Fanous, R.; Rivera-Hidalgo, F.; Binnie,; Hasegawa, S.; Lam, L.k.T. *Carcinogenesis*, **1989**, *10*, 1535.
24. Miller, E.G., Gonzales-Sanders, A.P.; Couvillon, A. M., Wright, J.M.; Hasegawa, S.; Lam, L.K.T. *Nutrition and Cancer* **1992** (in press)
25. Klocke, J. A.; Kubo,I. *Ent. Exp. & Appl.* **1982**, *32*,299.
26. Alford, A. R.; Bentley, M.D. *J. Econ. Entomol.* 1986,79,35.
27. Alford, A.R.; Cullen, J.A.; Storch, R.H.; Bentley, M.D.*J.Econ.Entomol.* **1987**, *80*, 575.
28. Arnason, J.T.; Philogene, B.J.R.; Donskov, D.; Kubo, I *Ent. Exp. & Appl.* **1987**,*43*:221.
29. Bentley, M.D.; Rajab, M.S.; Alford, A.R.; Mendel, M.J.; Hassanali, A. *Ent. Exp. & Appl.* **1989**, *49*, 189.
30. Bentley, M.D.; Rajab, M.S. Mendel, M.J.; Alford, A.R. *J. Agric. Food Chem.* **1990**, *38*, 1400.
31. Liu, Y.B.; Alford, A.R.; Rajab,R.S.; Bentley, M.D. *Physiol. Entomol.* **1990**, *15*, 37.

RECEIVED December 18, 1991

Chapter 9

Oxygenated C_{13}-Norisoprenoids
Important Flavor Precursors

Peter Winterhalter

Lehrstuhl für Lebensmittelchemie, Universität Würzburg, Am Hubland, D–8700 Würzburg, Germany

This chapter describes up to date developments in the field of natural flavor formation from C_{13}-norisoprenoid progenitors. The major pathways leading to important C_{13}-flavor compounds, such as, e. g., isomeric theaspiranes, vitispiranes, edulans, as well as ß-damascenone are discussed. In addition, further carotenoid degradation products, obtained from the central part of the polyene chain, are reported as natural products for the first time. Based on these results, possible routes of carotenoid degradation in plants are outlined.

Thirteen-carbon norisoprenoids derived from carotenoid degradation are common constituents in numerous plants (*1-4*). A great number of these compounds are important contributors to the overall fruit flavor and several of them also possess a considerable importance for the flavor and fragrance industry (*5-6*). Recent studies have shown that most of the volatile C_{13}-compounds identified so far are not genuine constituents of fruits but rather are derived from less or non-volatile precursor forms, such as polyhydroxylated norterpenoids as well as glycosidically bound progenitors. Since the first observation of the formation of a number of C_{13}-compounds as a result of heat-treatment of glycosidic wine and grape extracts by Williams *et al.* (*7*), considerable research interest has been directed to the identification of the corresponding non-volatile precursor forms. As a result of these efforts, the major pathways leading to the formation of important C_{13}-norisoprenoid compounds, such as, e.g., isomeric theaspiranes, vitispiranes, edulans as well as ß-damascenone, have been elucidated in recent years. A detailed description of the pathways giving rise to the formation of these highly esteemed C_{13}-odorants as well as biomimetic degradation reactions of selected carotenoid derived precursor compounds will be the focus of this chapter.

Studies on C_{13}-norisoprenoid precursors

The structures of the C_{13}-norisoprenoid precursors that have been the subject of our recent research interest are outlined in Figure 1. They comprise (i) mono- and diunsaturated C_{13}-diols (**1-9**), (ii) compounds with three oxygen functions (**10-12**) as well as (iii) acetylenic and allenic compounds (**13-16**). In the following, these compounds will be discussed with regard to their role as flavor precursors.

0097–6156/92/0490–0098$06.00/0
© 1992 American Chemical Society

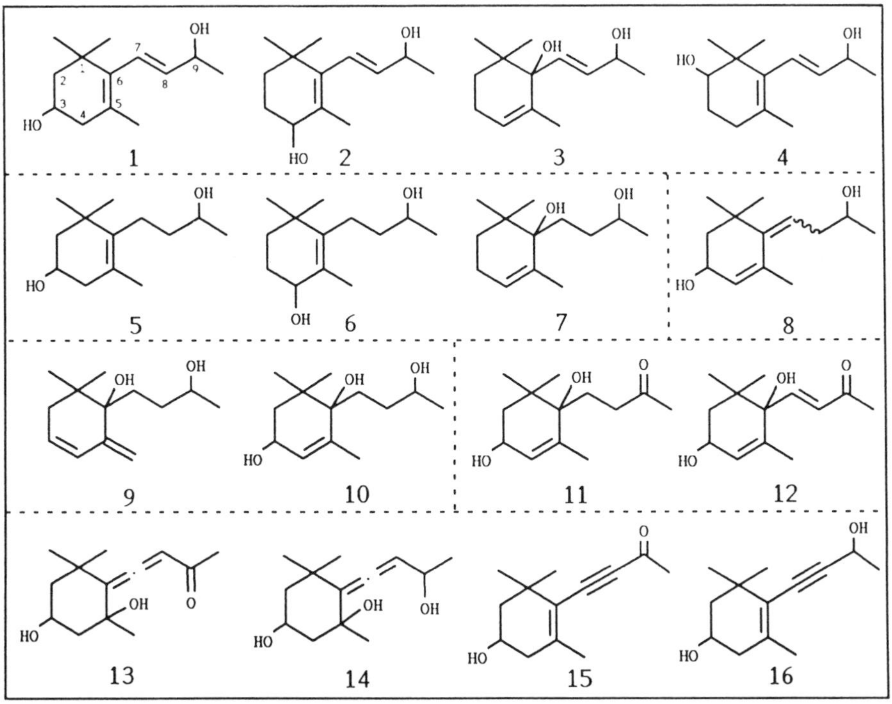

Figure 1.   Structures of important C$_{13}$-norisoprenoid precursor compounds: **1** 3-hydroxy-ß-ionol, **2** 4-hydroxy-ß-ionol, **3** 6-hydroxy-α-ionol, **4** 2-hydroxy-ß-ionol, **5** 3-hydroxy-7,8-dihydro-ß-ionol, **6** 4-hydroxy-7,8-dihydro-ß-ionol, **7** 6-hydroxy-7,8-dihydro-α-ionol, **8** 3-hydroxy-*retro*-α-ionol, **9** 6-hydroxy-3,4-dehydro-7,8-dihydro-γ-ionol, **10** 3,6-dihydroxy-7,8-dihydro-α-ionol, **11** 3,6-dihydroxy-7,8-dihydro-α-ionone, **12** 3,6-dihydroxy-α-ionone, **13** grasshopper ketone, **14** megastigma-6,7-diene-3,5,9-triol, **15** 3-hydroxy-7,8-dehydro-ß-ionone, **16** 3-hydroxy-7,8-dehydro-ß-ionol.

Hydrolytic chemistry of dienediols 1-4. The first group of precursor compounds with significant importance for "secondary" flavor formation in plants are the isomeric dienediols **1-4**.

Degradation of 3-hydroxy-ß-ionol 1. A recent study dealing with the influence of sample preparation on the composition of quince (*Cydonia oblonga* Mill.) flavor (8) revealed the formation of bicyclo [4.3.0] nonane derivatives **17** and **19** together with 3,4-didehydro-ß-ionol **18** (9,10) upon heat-treatment of quince juice at natural pH conditions (pH 3.5). As the volatiles **17-19** were absent in aroma extracts obtained by using gentle isolation techniques, i.e. high-vacuum distillation combined with liquid-liquid extraction, a generation of target compounds **17-19** from a non-volatile precursor form was evident. As a possible progenitor of **17-19** the known C_{13}-diol **1** (11,12) came under consideration. Initial research into the degradation of free diol **1** under simultaneous distillation/extraction (SDE) conditions, however, gave in addition to the degradation products outlined in Figure 2 further compounds bearing a hydroxyfunction at carbon-3 (13). The absence of such 3-hydroxyderivatives among the major quince norisoprenoids led to the conclusion that a glycosidically bound form of diol **1** might be responsible for the formation of quince volatiles **17-19**. It was assumed that a glycosidic linkage in the 3-position of diol **1** facilitates the introduction of the 3,4-double bond, thus yielding the pattern of volatiles outlined in Figure 2. Consequently, more intense studies into the glycosidic fraction of quince fruit became necessary. The glycosides were isolated by standard techniques, i.e. adsorption on XAD-resin followed by methanol-elution (14). After subsequent gentle fractionation of the glycosidic extract by rotation locular countercurrent chromatography (RLCC), peracetylation and a final purification by preparative HPLC, the precursor was identified as the ß-D-gentiobioside of 3-hydroxy-ß-ionol **1** with the sugar residue attached to carbon-3 (76). Model degradation studies carried out with this new glycoconjugate gave essentially the same pattern of volatiles as has been found in heat-treated quince juice. Quantitatively, considerable amounts of the disaccharide glycoside of **1**, i.e. approximately 10 mg per kg of fresh quince fruit, were determined.

Besides the major volatile degradation products **17-19**, a still unknown isomer of alcohol **19** as well as a whole series of C_{13}-hydrocarbons with similar mass spectral data as obtained for bicyclononatriene **17** (i.e. *m/z* 174, 159, 144) were also observed. One of these isomers was recently identified by Näf and Velluz (15) as triene **20**, which was isolated as a trace component from quince brandy, showing a "woody, camphoraceous and green" odor. The same group isolated also additional new flavor compounds from quince brandy, which are structurally related to bicyclo [4.3.0] nonane derivatives **17** and **19**, with most of them possessing interesting flavor properties (16).

The remaining, still unidentified degradation products of **1** with MS data similar to **17** consist of at least six further hydrocarbons with almost identical MS spectra. Based on MS- and FTIR-spectral data, these compounds were tentatively identified as isomeric tetraenes **21** and **22** (cf. Figure 2). Final confirmation of the proposed structures, however, must await the outcome of the chemical synthesis of the isomers, which failed so far, as the presumed tetraenes were found to be extremely unstable. Especially in the course of liquid chromatographic separation steps in dilute solution a rapid oxidation was observed, thus hampering the structural elucidation so far.

Concerning the occurrence of further conjugates of diol **1** in plants, unspecified bound forms of **1** were also detected in glycosidic extracts of papaya fruit (17) and, more recently, also in two *Prunus* species, i.e. apricot and peach (18). The identification of several degradation products of diol **1** in the volatile fraction of tomato paste (19) and starfruit (20) indicates that the C_{13}-flavor precursor **1** might obviously be more common in nature as previously known. Moreover, two glyco-

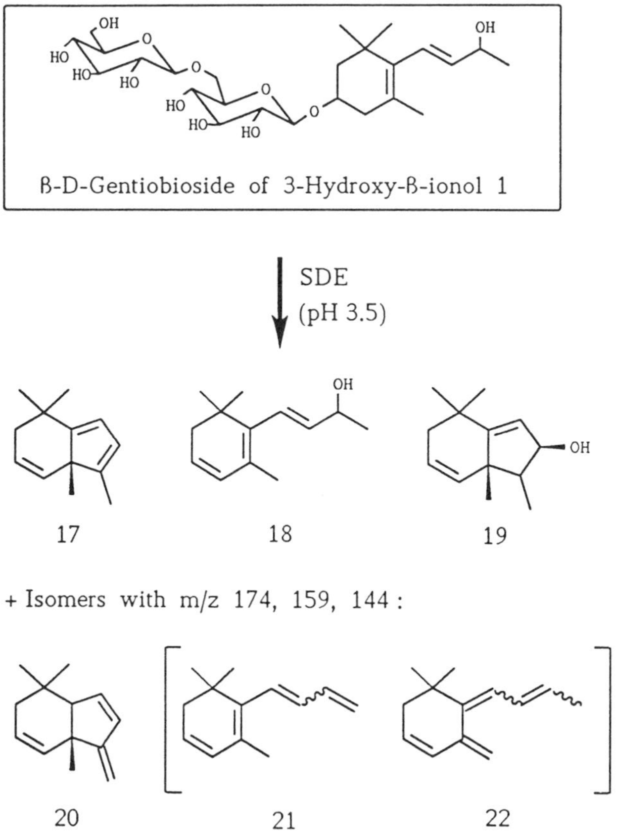

Figure 2.   Volatile degradation products obtained upon heat-treatment (SDE, pH 3.5) of glycosidically bound precursor **1**.

conjugates of an oxidized form of **1**, i.e. 3-hydroxy-ß-ionone, were also most recently isolated from quince fruit. These conjugates were elucidated as being the ß-D-gentiobioside as well as the ß-D-glucopyranoside of this particular norterpene (Güldner, A.; Winterhalter, P. *J. Agric. Food Chem.* **1991**, in press). In the case of the 3-hydroxyketone, it was possible to determine the absolute stereochemistry at carbon-3. The natural aglycon was found to be the (3*R*)-enantiomer, i.e. it had the same configuration as found in a series of ubiquitous carotenoids, such as, e.g., zeaxanthin. This finding can be regarded as a further support for the assumption that 3-hydroxy-ß-ionone and the related diol **1** are formed by oxidative degradation of corresponding 3-hydroxy-carotenoids.

Degradation of 4-hydroxy-ß-ionol 2. An increasing number of 4-oxygenated C_{13}-terpenoids has been identified during recent years (*21,22*). Among them, 4-hydroxy-ß-ionol **2** was elucidated as an important progenitor of a series of C_{13}-norisoprenoid aroma compounds. Diol **2** is a known constituent of quince fruit (*23*), tobacco (*24*) and *Osmanthus fragrans* (*21*). A nonspecified bound form of diol **2** was furthermore detected in purple passion fruit (*Passiflora edulis* Sims) (*25*). Biomimetic degradation reactions were carried out with free diol **2** under SDE-conditions (pH 3.5), using the modified Likens-Nickerson apparatus described by Schultz *et al.* (*26*). Under these conditions a rapid and complete degradation to the structures outlined in Figure 3 was observed. As major products (approx. 70 %) four isomeric megastigma-6,8-dien-4-ones **23 a-d** were obtained with the most abundant isomer, evaluated as the (6*E*,8*E*)-isomer by nOe-experiments, exhibiting a "pleasant weak tobacco note with a cooling effect" (*22*). Further volatiles included isomeric *retro*-α-ionones **24 a/b** as well as isomeric megastigma-5,8-dien-4-ones **25 a/b**. The latter isomers are well-known flavor compounds of *Osmanthus* absolute (*27*), yellow passion fruit (*Passiflora edulis* f. *flavicarpa*) and tobacco (*28*). The *E*-isomer **25a** has been described to exhibit a "fruity-floral and woody note". Minor degradation products of **2** were the hydrocarbon **17** and a still unidentified ketone (molecular weigth: 192), whose spectral data have been given earlier (*22*).

Separate degradation studies carried out with the structurally related diol **3** under SDE conditions gave less complete conversion to an identical pattern of volatiles as obtained for the 4-hydroxyderivative **2** (cf. Figure 3). This observation was explainable by the fact, that diol **3** itself was found to be steam volatile. Refluxing of diol **3** in a buffer solution (pH 3.5), however, led to its almost complete degradation - most probably *via* allylic rearrangement to the 4-hydroxyderivative **2** - as the same pattern of volatile degradation products as obtained for diol **2** was found. In nature, diol **3** is a rare compound. It was most recently seen for the first time as aglycon in a glycosidic extract of purple passion fruit (Herderich, M.; Winterhalter, P.; unpublished results).

Degradation of 2-hydroxy-ß-ionol 4. For completing the series of degradation studies on C_{13}-dienediols, the hydrolytic behavior of the 2-hydroxyderivative **4** at natural pH conditions was also investigated (cf. Figure 3). Contrary to ionone derivatives **1-3**, diol **4** has not yet been found as natural product. However, the oxidized form, i.e. 2-hydroxy-ß-ionone, is a known trace constituent of *Osmanthus* absolute (*21*) as well as of the edible algea *Porphyra tenera* (*29*). Moreover, 2-hydroxy-ß-ionone has been patented as tobacco flavorant (*30*).

Synthetic diol **4** was obtained by $LiAlH_4$-reduction of a donated sample of 2-hydroxy-ß-ionone. Diol **4** chromatographed with a Kováts index of 1652 on a J&W DB-5 capillary column and showed the following MS data (70eV): *m/z* (%) 192 (M^+-H_2O, 8), 177 (2), 174 (2), 159 (13), 149 (5), 121 (20), 119 (23), 91 (18), 69 (18), 55 (21), 43 (100). SDE treatment of **4** at pH 3.5 led to its complete degradation giving rise to a formation of four isomeric 2,5-epoxy megastigma-6,8-dienes **26 a-d** as main products (85 %), which are known odoriferous constituents of *Osmanthus*

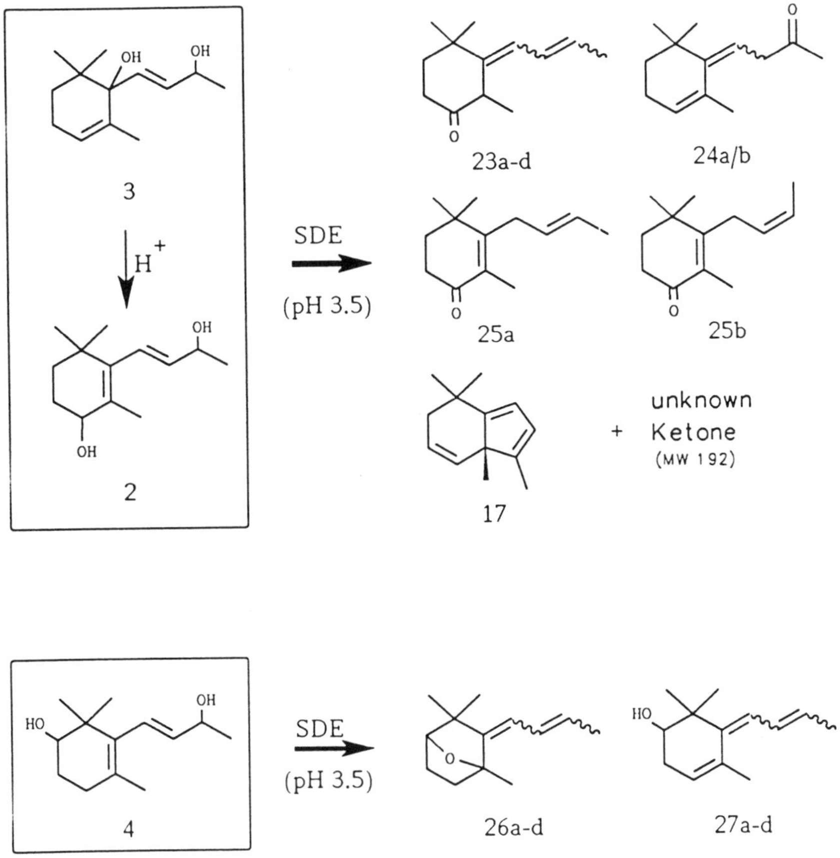

Figure 3. Top: Volatile degradation products obtained from 4-hydroxy-ß-ionol **2** (SDE, pH 3.5) as well as 6-hydroxy-α-ionol **3** (refluxing, pH 3.5). Bottom: Volatile degradation products formed after SDE-treatment (pH 3.5) of 2-hydroxy-ß-ionol **4**.

flowers (21). Especially the (6E,8E)-isomer is a useful flavoring substance; its odor has been described as "fresh and spicy, containing certain aspects of black currant buds, tomato leaves, and exotic fruit". Besides several not further studied trace components, isomeric hydroxytrienes **27 a-d** were observed as additional degradation products (13 %) of diol **4**.

Hydrolytic chemistry of enediols 5-7. The corresponding 7,8-dihydro analogues of dienediols **1-3** have been also found in nature. The hydrogenation of the side-chain double bond of diols **1-3** leads to a considerable change in the chemical reactivity. Whereas the 3-hydroxyderivative **5** turned out to be much less susceptible to acid treatment compared to the parent diol **1**, hydrogenation of 4-hydroxy-ß-ionol **2** generates a reactive 1,6-allyldiol system, which easily cyclizes under mild acidic conditions, giving rise to a formation of isomeric theaspiranes.

3-Hydroxy-7,8-dihydro-ß-ionol 5. Glycosidically bound forms of diol **5** have been detected in quince fruit (23) and, more recently, also in several *Prunus* species, i.e. apricot, peach, and yellow plum (18). Attempts to characterize the glycoconjugate of **5** from quince fruit, however, led only to a partial purification of the glycoside, as the sample still contained some ß-D-gentiobioside of diol **1** as impurity. From the thermospray mass spectra of the peracetylated glycoside, showing a strong pseudo-molecular ion at m/z 890 [872 + NH_4]$^+$, the presence of a disaccharide moiety was evident. Enzymatic hydrolysis using almond emulsin liberated the aglycon **5** and ß-D-glucose as sole sugar. Based on the MS data obtained as well as the result of emulsin treatment, it was concluded that diol **5** is - in analogy to the parent compound **1** - also present in quince fruit as ß-D-gentiobioside. Since in the acid hydrolysates (SDE, pH 3.5) of the deacetylated glycoside no volatiles derivable from aglycon **5** were detected, this compound is obviously not a source of volatile products in fruits.

Formation of isomeric theaspiranes from diol 6. Contrary to diol **5**, the 4-hydroxyisomer **6** is characterized by an extreme reactivity due to the 1,6-allyldiol structure. Even under mild reaction conditions (natural pH, room temperature) a cyclization to the corresponding tetrahydrofurane derivatives, i. e. isomeric theaspiranes, occurs (cf. Figure 4). Such cyclization reactions have been used first by Ohloff et al. (31) for the preparation of monoterpene volatiles. Among the C_{13}-norterpenes such a reaction, i.e. a "prototropic dehydration of 1,6-allyldiols" was observed by us for the first time. The natural theaspirane precursor **6** was initially isolated from a polar quince fruit extract (32); in the meantime a nonspecified bound form of **6** has also been detected in a glycosidic passion fruit isolate (25). Isomeric theaspiranes are highly esteemed by the flavor industry. In tobacco, an addition of only 5 ppb of theaspiranes suffice to reach the optimum flavor, for other products 1 ppb of the spiro ethers are used to round off the fruit aroma (33).

In addition to diol **6**, the structurally related diol **7** has been also identified recently as trace component in a polar quince fruit extract (Winterhalter, P.; Schreier, P.; unpublished results). 6-Hydroxy-7,8-dihydro-α-ionol **7** chromatographed on a J&W DB-5 capillary column with a Kováts index of 1607 and showed the following MS data (70 eV): m/z (%) 194 (M^+-H_2O, 4), 179 (1), 176 (1), 156 (17), 138 (92), 119 (26), 109 (27), 96 (49), 95 (39), 82 (63), 67 (16), 55 (41), 43 (100), 41 (46). Synthesis of **7**, attempted by hydrogenation of diene diol **3**, gave the desired product **7** only in a low yield, therefore biomimetic degradation reactions with **7** have not been carried out so far. However, under acidic conditions a formation of isomeric theaspiranes - *via* allylic rearrangement to diol **6** - may be expected. An alternative synthetic access to diol **7** has recently been published by Weyerstahl et al. (34).

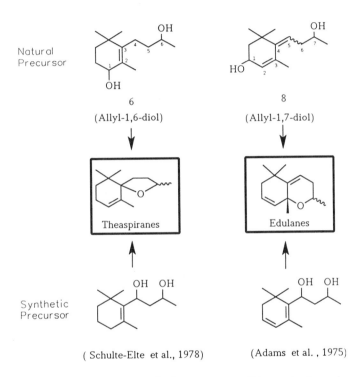

Figure 4. Natural as well as synthetic precursors of isomeric theaspiranes and isomeric edulanes.

Identification of 3-hydroxy-*retro*-α-ionol as a natural precursor of isomeric edulans. Isomeric edulans (cf. Figure 4) are reported to be characterized by an "attractive roselike odor", and they are considered as key flavor compounds of purple passion fruit (*35-38*). Synthetic progenitors of isomeric edulans are well-known (*39-41*), a natural precursor, however, has only been identified most recently by our group (*42*). As a first step in the study on edulan genesis, it was necessary to examine whether the target compounds are formed by degradation of glycosidically bound constituents or by an acid-catalyzed cyclization of free polyols. As the edulans were absent in acid-hydrolysates (SDE, pH 3.5) of glycosidic passion fruit extracts, our attention was directed to the polyol fraction. In analogy to theaspirane formation from 1,6-allyldiol **6**, isomeric edulans should theoretically be accessible *via* prototropic dehydration and subsequent cyclization of the previously unknown 1,7-allyldiol **8**. Precedent for this type of reaction can again be found in earlier work on the dehydration of monoterpenoid 1,7-allyldiols, giving rise to a formation of tetrahydropyrane derivatives, such as diastereoisomeric rose oxides as well as nerol oxide, respectively (*31,43*). The presumed precursor **8** was synthezised by isomerization of 3-oxo-α-ionol to the corresponding *retro*-compounds, followed by mild reduction using $NaAlH_2(OCH_2CH_2OCH_3)_2$ (*44*). With the so-obtained reference compounds, the (*E*)-isomer of diol **8** was identified for the first time in purple passion fruit (*42*). Degradation reaction carried out with **8** proved the easy formation of isomeric edulans under acidic conditions.

Investigation of the formation of C_{13}-norisoprenoid volatiles in Riesling wine. Among the C_{13}-norisoprenoid odorants present in aged Riesling wine, three components, i.e. isomeric vitispiranes, 1,1,6-trimethyl-1,2-dihydronaphthalene (TDN), and ß-damascenone, are regarded as important contributors to the bottle-aged bouquet (*45-47*). TDN, in particular, with a low sensory threshold (20 ppb in wine medium) was found to be responsible for a "hydrocarbon or kerosene-like note" present in aged Riesling wines (*48*). Initial research into the origin of these norisoprenoids indicated that they are formed by a degradation of glycosidically bound progenitors (*7,49*). However, due to the structural diversity of the numerous bound constituents present in wine, progress in the separation and purification of individual precursor compounds has been slow. It was only in recent years that the all-liquid chromatographic technique of countercurrent chromatography has been recognized to be of paramount importance for the fractionation of the complex glycosidic mixture of wine and grapes (*50*). Especially, droplet countercurrent chromatography (DCCC) has been employed for the preparative separation of glycosidic isolates from grape juice and wine. Application of the new technique of two-dimensional GC-DCCC analysis on Riesling glycosides revealed the presence of almost one hundred - various glycosylated - aglycons in this particular wine (*51*). The same technique has then also been employed for the separation of the precursors of vitispirane, TDN and ß-damascenone. As a result of DCCC prefractionation and subsequent acid hydrolyses (SDE, pH 3.2) of separated DCCC-fractions, it was concluded that multiple precursors are responsible for the formation of these important C_{13}-flavor compounds in wine (*52*).

Formation of isomeric vitispiranes (cf. Figure 5). Those fractions that generated isomeric vitispiranes upon heat-treatment were found to contain unspecified bound forms of triol **30** as well as isomeric hydroxytheaspiranes **31**. Assuming that the glycosidic linkage is located in the 3-position, a cyclization in analogy to theaspirane formation can take place. In addition to the aglycons **30** and **31**, we were most recently successful in identifying a further vitispirane-yielding precursor. In enzymatic hydrolysates of separated DCCC-fractions, two diastereoisomers of dienediol **9**, which are known synthetic progenitors of the

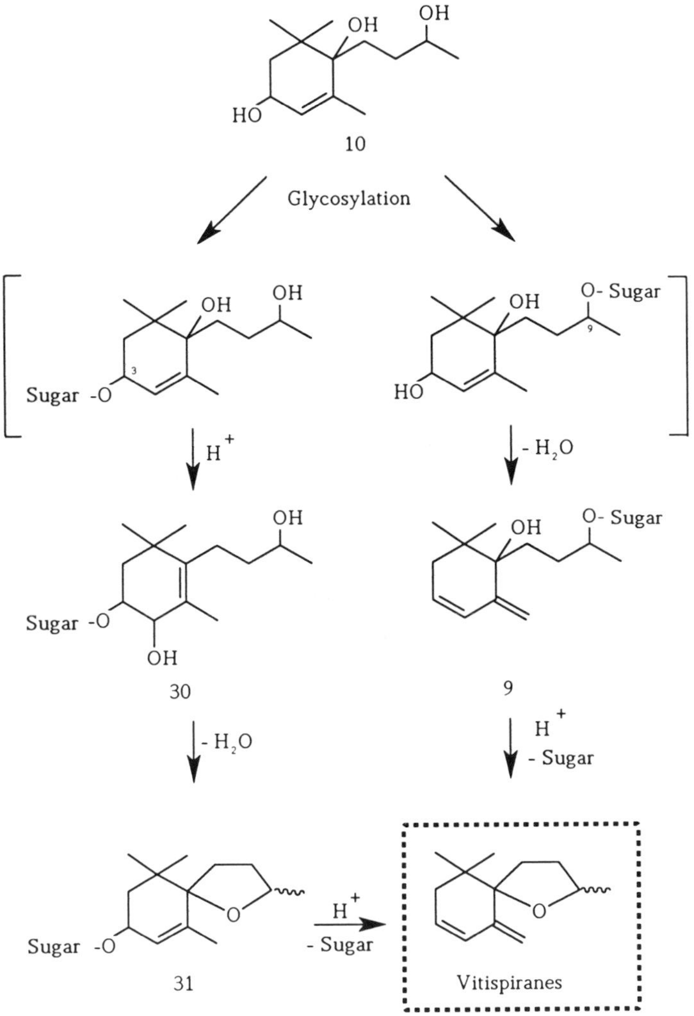

Figure 5. Proposed pathway of vitispirane formation in Riesling wine.

odoriferous spiroethers (53), were observed by us for the first time as natural products (Waldmann, D.; Winterhalter, P. *Vitis* **1991**, submitted). With the finding of the additional vitispirane-yielding aglycon **9** the following pathway for vitispirane formation can be proposed (cf. Figure 5). As initial biogenetic precursor triol **10** can be regarded, which after glycosylation in the 3- or 9-position is expected to react in two different ways: First, with the sugar attached to the 3-position, the conjugate is susceptible to an acid induced allylic rearrangement giving rise to a formation of the thermodynamically more stable isomer **30** as previously shown by Strauss *et al.* (54). Second, with the glycosidic linkage in the side-chain a reactive hydroxyfunction in the 3-position is left, which after dehydration generates the new natural vitispirane precursor **9**.

Formation of 1,1,6-trimethyl-1,2-dihydronaphthalene (TDN) in Riesling wine (cf. Figure 6). Not only triol **10** seems to be important for "secondary" flavor formation in wine, but also studies on the hydrolysis of the oxidized form, i.e. 3,6-dihydroxy-7,8-dihydro-α-ionone **11a**, gave a further important insight into the hydrolytic processes occurring during the storage of wine. Initially, compound **11a** was used as a synthetic progenitor of the so-called "Riesling acetal" **33**, a new aroma compound of Riesling wine (55), which was independently also found in quince brandy by Näf *et al.* (56). The equilibrium product of **11a**, hemiacetal **11b**, represents a further example for a reactive 1,7-allyldiol. It was therefore assumed to generate the intramolecular acetal **33** by an acid-catalyzed cyclization. A biomimetic degradation of **11b** was carried under SDE conditions at pH 3.2, and Riesling acetal **33** was indeed obtained as major volatile product (in approximately 60 % yield). Besides two minor volatiles, i.e. spiroether **32** and the aromatic ketone **34**, the off-flavor causing wine constituent TDN **36** and a hydroxylated derivative of TDN were found as further degradation products of hemiacetal **11b**. The hydroxyderivative of TDN has been seen earlier in aged Riesling wine by Di Stefano (57), who tentatively assigned the structure as 1,1,6-trimethyl-4-hydroxy-1,2,3,4-tetrahydronaphthalene. However, from the vapor-phase FTIR spectrum obtained for this compound a nonaromatic structure is apparent. Furthermore, FTIR data revealed the presence of a tertiary hydroxygroup as well as a strong absorption band at 733 cm^{-1}, being characteristic for cis-configurated double bonds. These data are in accordance with the proposed structure **35**, whose synthesis is presently in progress.

An identical pattern of volatiles - as obtained in the degradation reactions carried out with synthetic hemiacetal **11b** - has also been found in acid-hydrolysates of separated DCCC fractions of Riesling glycosides. This indicated already indirectly the presence of several glycoconjugated forms of **11b** in this wine. With trimethylsilylated references of precursors **11** in hand, we succeeded finally in identifying hemiacetal **11b** in the aglycon fraction of Riesling wine (Winterhalter, P. *J. Agric. Food Chem.* **1991**, in press). Moreover, the allylic rearranged isomer **11c** could also be identified as an additional Riesling aglycon, which was found to give the same pattern of volatile degradation products upon heat-treatment at pH 3.2 as obtained for hemiacetal **11b** (Humpf, H.U.; Winterhalter, P.; Schreier, P. *J. Agric. Food Chem.* **1991**, in press). The elucidation of the entire structure of the different TDN-yielding glycosides remains the subject of continuing research.

The finding of isomeric compounds **11** as Riesling aglycons represents to the authors best knowledge the first identification of major TDN-yielding precursors in wine. Although these results will not prevent the development of the TDN-caused 'hydrocarbon or kerosene off-flavor' in wine, we slowly start to understand the pathways by which the off-flavor is formed. The knowledge about the structures and the hydrolysis of TDN precursors will hopefully help the wine industry in the future to select wines that are more stable during prolonged bottle storage.

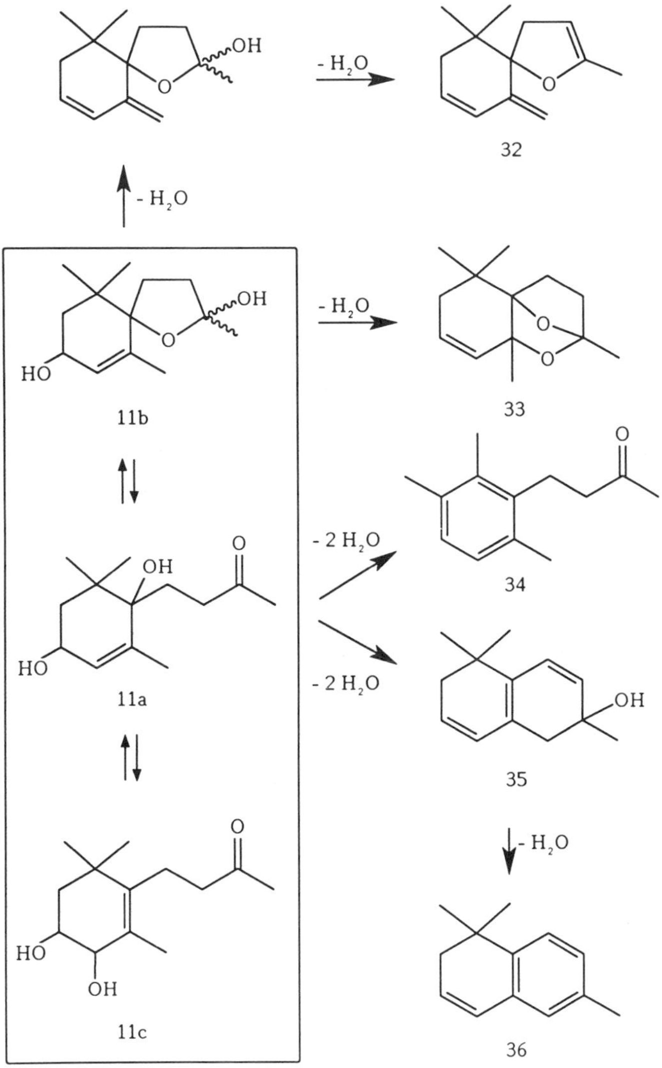

Figure 6. Structures of the newly identified natural TDN-precursors **11b** and **11c** as well as reaction steps rationalizing the formation of the Riesling wine volatiles **32-36** (adapted from Winterhalter, P., *J. Agric. Food Chem.* **1991**, in press).

Stimulated by the results obtained for compound **11**, we were also interested in the hydrolytic chemistry of the 7,8-unsaturated derivative **12**. However, in the case of 3,6-dihydroxy-α-ionone **12** only two volatile degradation products were observed upon heat-treatment employing SDE at pH 3.5. Simple dehydration yielded 3,4-dehydro-6-hydroxy-γ-ionone, and the loss of two molecules of water resulted in the formation of 4-(2',3',6'-trimethyl)-but-3-en-2-one. The latter hydrocarbon is a known flavor compound of yellow passionfruit, exhibiting "green, fruity, and woody notes" (58).

Formation of ß-damascenone (cf. Figure 7). Among the approximately 5000 known volatile flavoring substances the C_{13}-norisoprenoid ketone, ß-damascenone, is clearly one of the most outstanding. Since its discovery in rose oil (*Rosa damascena*) by Demole *et al.* (59) it greatly influenced present-day flavor and fragrance chemistry. ß-Damascenone has an odor threshold of 0.002 in water (60), making it one of the most potent of all known wine flavor compounds. The observation of ß-damascenone in acid-hydrolysates of grape juice and wine (7,61) indicated again a formation from non-volatile precursor compounds. However, contrary to the majority of the C_{13}-norisoprenoids isolated to date, most of them bearing an oxygen-function in the side-chain at C-9, ß-damascenone is a 7-oxygenated norterpenoid. Consequently, simple carotenoid degradation alone cannot generate this structure.

In the early work of Ohloff *et al.* (62) and Isoe *et al.* (63) the question of oxygen transposition from C-9 to C-7 was for the first time explained by the assumption of allenic intermediates formed by biodegradation of allenic carotenoids, such as neoxanthin. As a primary degradation product, the so-called grasshopper ketone **13** was regarded. Enzymatic reduction would then give the allenic triol **14**, which - on the basis of biomimetic studies on structurally related compounds - was expected to yield the target compound, ß-damascenone, together with 3-hydroxy-ß-damascone under acidic conditions. Most recently, such allenes have been detected in natural substrates. A disaccharide glycoside of **13** has been isolated from *Cinnamomum cassia* (64), and the ß-D-glucopyranoside of the allenic triol **14** was isolated from *Lycium halimifolium* (65). Unspecified bound forms of allenes **13** and **14** have also been detected in wine (51,66), where they are present in trace amounts.

In addition to such allenic progenitors, acetylenic intermediates seem also to be involved in ß-damascenone formation. Acetylenic diol **16** has first been isolated from tobacco (67), in the meantime **16** has been additionally found as aglycon in wine (68) as well as purple passion fruit (69). Diol **16** is clearly a logical intermediate in the pathway proposed for ß-damascenone formation from allenic progenitors. However, in an increasing number of substrates we were also able to detect the oxidized form, the acetylenic hydroxyketone **15**. Especially with the help of on-line HRGC-FTIR spectroscopy a fast screening of plant extracts for such acetylenes was possible. The frequent identification of ketone **15** in an increasing number of plant tissues (69) indicates that besides allenic carotenoids and intermediates, acetylenic carotenoids should also be considered as initial ß-damascenone yielding progenitors.

Biodegradation of carotenoids.

C_{13}-norisoprenoids - like, e. g., ß-ionone - are generally regarded as carotenoid degradation products (70), resulting from an oxidative cleavage of the polyene chain in the 9,10-position (cf. Figure 8). Additional degradation products consist of 9-, 10- or 11-carbon norisoprenoids, and even for C_{15}-compounds like the important plant hormone, abscisic acid, an 'apo-carotenoid' pathway has been suggested (71). But this raises the question - what happens with the rest of the polyene chain after the

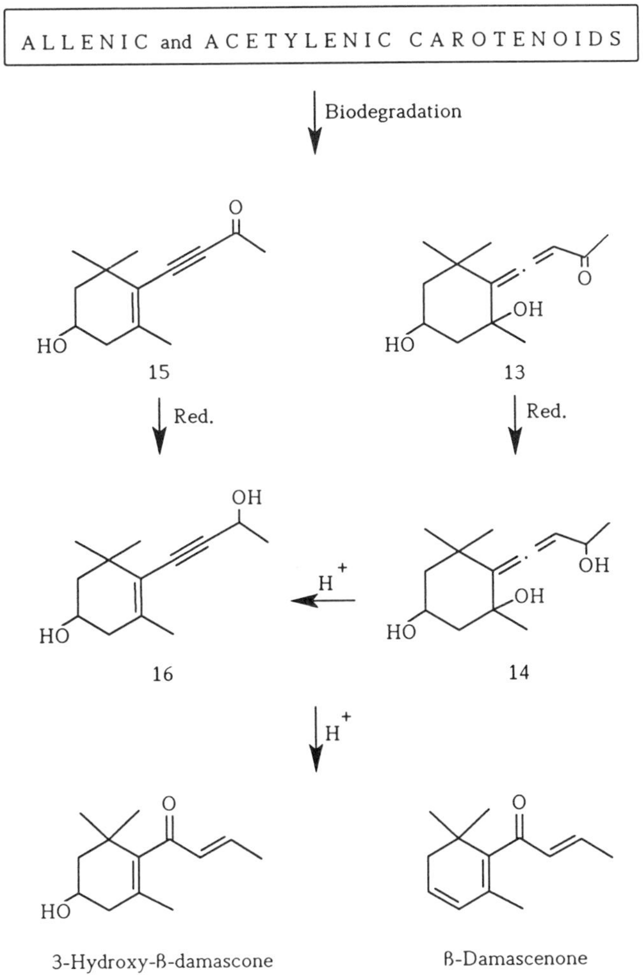

Figure 7. Proposed pathway for ß-damascenone formation from allenic and acetylenic carotenoids.

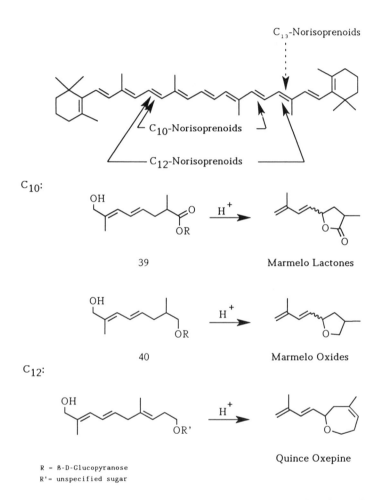

Figure 8. Biodegradation of carotenoids resulting in the formation of C_{13}-norisoprenoids as well as C_{10} and C_{12}-terpenoids as important cleavage products.

cleavage of the endgroups? Most recenty, we succeeded in identifying glycoconjugated irregular monoterpenoids as progenitors of isomeric marmelo lactones and marmelo oxides, which are considered as key flavor compounds of quince fruit (*72,73*). The ß-D-glucopyranoside of 2,7-dimethyl-8-hydroxy-4(*E*),6(*E*)-octadienoic acid **39** was identified as a natural precursor of isomeric marmelolactones (*74*), and the ß-D-glucopyranoside of the corresponding diol **40** as precursor of isomeric marmelo oxides (Lutz, A.; Winterhalter, P.; Schreier, P. *Tetrahedron Lett.* **1991**, in press). In addition, Näf and Velluz (*75*) isolated a C_{12}-norterpene as precursor of quince oxepines. Such norterpenoid compounds can be regarded as missing links in carotenoid degradation. They are obviously derived from the central part of the polyene chain - which is left after the cleavage of the end-groups (cf. Figure 8). Moreover, the finding of **39** and **40** shows that also the central part of carotenoids can serve as important flavor precursors.

Summarizing, it is obvious that during recent years considerable progress in the field of C_{13}-norisoprenoids has been made. As a result, the major pathways giving rise to a formation of important C_{13}-odorants, such as e.g. ß-damascenone, theaspiranes, vitispiranes, and edulans from nonvolatile precursors could be elucidated. In spite of these efforts, remarkably little is known about the initial steps of carotenoid degradation and the subsequent enzymatic pathways leading to the flavor precursors that have been discussed in this chapter. These still open questions will remain interesting research fields for the future.

Acknowledgments. Dr. D. Lamparsky, Givaudan, is thanked for a generous sample of 2-hydroxy-ß-ionone. The Deutsche Forschungsgemeinschaft, Bonn, is thanked for funding the research.

Literature cited

1. Ohloff, G.; Flament, I.; Pickenhagen, W. *Food Rev. Int.* **1985**, *1*, 99-148.
2. Wahlberg, I.; Enzell, C.R. *Nat. Prod. Rep.* **1987**, *4*, 237-276.
3. Winterhalter, P.; Schreier, P. In *Bioflavour'87*; Schreier, P., Ed.; De Gruyter: Berlin, New York, 1988; pp 255-273.
4. Williams, P.J.; Sefton, M.A.; Wilson, B. In *Flavor Chemistry - Trends and Developments*; Teranishi, R.; Buttery, R.G.; Shahidi, F., Eds.; ASC Symp. Ser. 388; American Chemical Society: Washington, DC, 1989; pp 35-48.
5. Ohloff, G. *Riechstoffe und Geruchssinn - Die molekulare Welt der Düfte*; Springer Verlag: Berlin, 1990.
6. Ohloff, G.; Demole, E. *J. Chromatogr.* **1987**, *406*, 181-183.
7. Williams, P.J.; Strauss, C.R.; Wilson, B.; Massy-Westropp, R.A. *J. Chromatogr.* **1982**, *235*, 471-480.
8. Winterhalter, P.; Lander, V.; Schreier, P. *J. Agric. Food Chem.* **1987**, *35*, 335-337.
9. Tsuneya, T.; Ishihara, M.; Shiota, H.; Shiga, M. *Agric. Biol. Chem.* **1983**, *47*, 2495-2502.
10. Ishihara, M.; Tsuneya, T.; Shiota, H.; Shiga, M.; Nakatsu, K. *J. Org. Chem.* **1986**, *51*, 491-495.
11. Fujimori, T.; Kasuga, R.; Kaneko, H.; Noguchi, M. *Agric. Biol. Chem.* **1975**, *39*, 913-914.
12. Sannai, A.; Fujimori, T.; Uegaki, R.; Akaki, T. *Agric. Biol. Chem.* **1984**, *48*, 1629-1630.
13. Winterhalter, P.; Schreier, P. In *Thermal Generation of Aromas*; Parliment, T.H.; Ho, C.T.; McGorrin, R.J., Eds.; ASC Symp. Ser. 409; American Chemical Society: Washington, DC, 1989; pp 320-330.
14. Günata, Y.Z.; Bayonove, C.L.; Baumes, R.L.; Cordonnier, R.E. *J. Chromatogr.* **1985**, *331*, 83-90.

15. Näf, R.; Velluz, A. *J. Ess. Oil Res.* **1991**, *3*, 165-172.
16. Näf, F.; Näf, R.; Uhde, G. In *Flavour Science and Technology*; Bessière, Y.; Thomas, A.F., Eds.; Wiley & Sons: Chichester, New York, 1990; pp 3-20.
17. Schwab, W.; Mahr, C.; Schreier, P. *J. Agric. Food Chem.* **1989**, *37*, 1009-1012.
18. Krammer, G.; Winterhalter, P.; Schwab, M.; Schreier, P. *J. Agric. Food Chem.* **1991**, *39*, 778-781.
19. Buttery, R.G.; Teranishi, R.; Flath, R.A.; Ling, L.C. *J. Agric. Food Chem.* **1990**, *38*, 792-795.
20. MacLeod, G.; Ames, J.M. *Phytochemistry* **1990**, *29*, 165-172.
21. Kaiser, R.; Lamparsky, D. *Helv. Chim. Acta* **1978**, *61*, 373-382.
22. Winterhalter, P.; Herderich, M.; Schreier, P. *J. Agric. Food Chem.* **1990**, *38*, 796-799.
23. Winterhalter, P.; Schreier, P. *J. Agric. Food Chem.* **1988**, *36*, 1251-1256.
24. Weeks, W.W.; Seltmann, H. *J. Agric. Food Chem.* **1986**, *34*, 899-904.
25. Winterhalter, P. *J. Agric. Food Chem.* **1990**, *38*, 452-455.
26. Schultz, T.H.; Flath, R.A.; Mon, T.R.; Eggling, S.B.; Teranishi, R. *J. Agric. Food Chem.* **1977**, *25*, 446-449.
27. Kaiser, R.; Lamparsky, D. *Helv. Chim. Acta* **1978**, *61*, 2328-2335.
28. Demole, E.; Enggist, P.; Winter, M.; Furrer, A.; Schulte-Elte, K.H.; Egger, B.; Ohloff, G. *Helv. Chim. Acta* **1979**, *62*, 67-75.
29. Flament, I.; Ohloff, G. In *Progress in Flavour Research 1984*; Adda, J., Ed.; Elsevier: Amsterdam, 1985; pp 281-300.
30. Jpn. Patent *78,107,497* **1978**; *Chem. Abstr.* **1979**, *90*, 19280w.
31. Ohloff, G.; Schulte-Elte, K.-H.; Willhalm, B. *Helv. Chim. Acta* **1964**, *47*, 602-626.
32. Winterhalter, P.; Schreier, P. *J. Agric. Food Chem.* **1988**, *36*, 560-562.
33. Ger. Offen. *2,610,238* **1976**; *Chem. Abstr.* **1977**, *86*, 95876c.
34. Weyerstahl, P.; Buchmann, B.; Marschall-Weyerstahl, H. *Liebigs Ann. Chem.* **1988**, 507-523.
35. Whitfield, F.B.; Stanley, G.; Murray, K.E. *Tetrahedron Lett.* **1973**, 95-98.
36. Whitfield, F.B.; Stanley, G. *Aust. J. Chem.* **1977**, *30*, 1073-1091.
37. Adams, D.R.; Bhatnagar, S.P.; Cookson, R.C.; Stanley, G.; Whitfield, F.B. *J. Chem. Soc., Chem. Commun.* **1974**, 469-470.
38. Whitfield, F.B.; Last, J.H. In *Progress in Essential Oil Research*; Brunke, E.J., Ed.; de Gruyter: Berlin, New York, 1986; pp 3-48.
39. Adams, D.R.; Bhatnagar, S.P.; Cookson, R.C. *J. Chem. Soc., Perkin Trans. 1* **1975**, 1736-1739.
40. Schulte-Elte, K.H.; Gautschi, F.; Renold, W.; Hauser, A.; Fankhauser, P.; Limacher, J.; Ohloff, G. *Helv. Chim. Acta* **1978**, *61*, 1125-1133.
41. Etoh, H.; Ina, K.; Iguchi, M. *Agric. Biol. Chem.* **1980**, *44*, 2871-2876.
42. Herderich, M.; Winterhalter, P. *J. Agric. Food Chem.* **1991**, *39*, 1270-1274.
43. Ohloff, G.; Lienhard, B. *Helv. Chim. Acta* **1964**, *48*, 182-189.
44. Prestwich, G.D.; Whitfield, F.B.; Stanley, G. *Tetrahedron* **1976**, *32*, 2945-2948.
45. Simpson, R.F.; Miller, G.C. *Vitis* **1983**, *22*, 51-63.
46. Strauss, C.R.; Wilson, B.; Anderson, R.; Williams, P.J. *Am. J. Enol. Vitic.* **1987**, *38*, 23-27.
47. Rapp, A.; Güntert, M.; Ullemeyer, H. *Z. Lebensm. Unters. Forsch.* **1985**, *180*, 109-116.
48. Simpson, R.F. *Chem. & Ind.* **1978**, 37.
49. Williams, P.J.; Strauss, C.R.; Wilson, B.; Dimitriadis, E. In *Topics in Flavour Research*, Berger, R.G.; Nitz, S.; Schreier, P., Eds.; Marzling, FRG: Eichhorn, 1985; pp 335-352.

50. Strauss, C.R.; Gooley, P.R.; Wilson, B.; Williams, P.J. *J. Agric. Food Chem.* **1987**, *35*, 519-524.
51. Winterhalter, P.; Sefton, M.A.; Williams, P.J. *J. Agric. Food Chem.* **1990**, *38*, 1041-1048.
52. Winterhalter, P.; Sefton, M.A.; Williams, P.J. *Am. J. Enol. Vitic.* **1990**, *41*, 277-283.
53. Kato, T.; Kondo, H. *Bull. Chem. Soc. Jpn.* **1981**, *54*, 1573-1574.
54. Strauss, C.R.; Dimitriadis, E.; Wilson, B.; Williams, P.J. *J. Agric. Food Chem.* **1986**, *34*, 145-149.
55. Winterhalter, P.; Sefton, M.A.; Williams, P.J. *Chem. & Ind.* **1990**, 463-464.
56. Näf, R.; Velluz, A.; Decorzant, R.; Näf, F. *Tetrahedron Lett.* **1991**, *32*, 753-756.
57. Di Stefano, R. *Riv. Vitic. Enol.* **1985**, *38*, 228-241.
58. Ger. Offen. 2,445,649 **1975**; *Chem. Abstr.* **1975**, *83*, 43050x.
59. Demole, E.; Enggist, P.; Säuberli, U.; Stoll, M.; Kováts, E.sz. *Helv. Chim. Acta* **1970**, *53*, 541-551.
60. Buttery, R.G.; Teranishi, R.; Ling, L.C. *Chem. & Ind.* **1988**, 238.
61. Braell, P.A.; Acree, T.E.; Butts, R.M.; Zhou, P.G. In *Biogeneration of Aromas*, Parliment, T.H.; Croteau, R., Eds.; ASC Symp. Ser. 317; American Chemical Society: Washington, DC, 1986; pp 75-84.
62. Ohloff, G.; Rautenstrauch, V.; Schulte-Elte, K.H. *Helv. Chim. Acta* **1973**, *56*, 1503-1513.
63. Isoe, S.; Katsumura, S.; Sakan, T. *Helv. Chim. Acta* **1973**, *56*, 1514-1516.
64. Shiraga, Y.; Okano, K.; Akira, T.; Fukaya, C.; Yokoyama, K.; Tanaka, S.; Fukui, H.; Tabata, M. *Tetrahedron* **1988**, *44*, 4703-4711.
65. Näf, R.; Velluz, A.; Thommen, W. *Tetrahedron Lett.* **1990**, *31*, 6521-6522.
66. Skouroumounis, G.K. *Ph. D. Thesis*, University of Adelaide, South Australia, 1991.
67. Fujimori, T.; Kasuga, R.; Kaneko, H.; Noguchi, M. *Phytochemistry* **1975**, *14*, 2095.
68. Sefton, M.A.; Skouroumounis, G.K.; Massy-Westropp, R.A.; Williams, P.J. *Aust. J. Chem.* **1989**, *42*, 2071-2084.
69. Winterhalter, P.; Full, G.; Herderich, M.; Schreier, P. *Phytochem. Anal.* **1991**, *2*, 93-96.
70. Enzell, C. *Pure & Appl. Chem.* **1985**, *57*, 693-700.
71. Parry, A.D.; Horgan, R. *Phytochemistry* **1991**, *30*, 815-821.
72. Tsuneya, T.; Ishihara, M.; Shiota, H.; Shiga, M. *Agric. Biol. Chem.* **1980**, *44*, 957-958.
73. Nishida, Y.; Ohrui, H.; Meguro, H. *Agric. Biol. Chem.* **1983**, *47*, 2969-2971.
74. Winterhalter, P.; Lutz, A.; Schreier, P. *Tetrahedron Lett.* **1991**, *32*, 3669-3670.
75. Näf, R.; Velluz, A. *Tetrahedron Lett.* **1991**, *32*, 4487-4490.
76. Winterhalter, P.; Harmsen, S.; Trani, F. *Phytochemistry* **1991**, *30*, 3021-3025.

RECEIVED December 18, 1991

Chapter 10

Free and Bound Flavor Constituents of White-Fleshed Nectarines

Gary R. Takeoka[1], Robert A. Flath[1], Ron G. Buttery[1], Peter Winterhalter[2], Matthias Güntert[3], David W. Ramming[4], and Roy Teranishi[1]

[1]Agricultural Research Service, U.S. Department of Agriculture, 800 Buchanan Street, Albany, CA 94710
[2]Lehrstuhl für Lebensmittelchemie, Universität Würzburg, Am Hubland, D–8700 Würzburg, Germany
[3]Research Department, Haarmann and Reimer GmbH, Postfach 1253, D–3450 Holzminden, Germany
[4]Agricultural Research Service, U.S. Department of Agriculture, 2021 South Peach Avenue, Fresno, CA 93727

The volatiles of white-fleshed nectarine (*Prunus persica nectarina*) were studied by capillary gas chromatography and combined capillary gas chromatography-mass spectrometry (GC-MS) using dynamic headspace sampling of intact fruit, direct extraction of juice, and simultaneous vacuum and atmospheric distillation-extraction of blended fruit. Enzymatic hydrolysis of the glycoside fraction obtained from nectarine juice by adsorption on Amberlite XAD-2 resin followed by elution with ethyl acetate and methanol led to the identification of the following C_{13} norisoprenoids: 3-oxo-α-ionol, 3-hydroxy-7,8-dihydro-β-ionol, 3-hydroxy-β-ionone, 3-oxo-7,8-dihydro-α-ionol, 3-oxo-retro-α-ionol, vomifoliol, and 7,8-dihydrovomifoliol.

There have only been limited studies on the volatile constituents of nectarine (*1,2*). While the flavor contribution of lactones in nectarines has been clarified (*1*), flavor differences between white and yellow-fleshed nectarines have yet to be explained. White-fleshed nectarines typically possess an additional flowery note which is lacking in the yellow-fleshed varieties. One goal of this study was to determine what compounds are responsible for this unique odor characteristic in white-fleshed nectarines.

0097–6156/92/0490–0116$06.75/0
© 1992 American Chemical Society

The precursor analysis approach has been demonstrated to be a useful complement to conventional methods of fruit flavor research (3). Recent studies on various *Prunus* species such as apricot, peach, yellow plum (4) and sour cherry (5) have shown that these fruits contain conjugated forms of monoterpenes, C13 norisoprenoids and shikimic-acid derived metabolites and that these conjugates play an important role as flavor precursors. This work reports on the flavor constituents liberated and produced during enzymatic and acid hydrolysis of the glycoside fraction of white-fleshed nectarines.

Experimental

Materials. Fresh tree-ripened nectarines (*Prunus persica nectarina*), cultivar P 89-56, were obtained from orchards of the Horticultural Crops Research Laboratory of the U.S. Department of Agriculture, Fresno, CA. P 89-56 is a white-fleshed experimental cultivar being tested in Fresno, CA and was grown under standard cultivation methods with flood irrigation.

Sample Preparation. 1. Dynamic Headspace Sampling. Intact fruit (total weight 3.9 kg) were placed in a 9-L Pyrex glass container. A Pyrex head was attached to the top of the container which allowed purified air to enter the bottom of the chamber (via a Teflon tube) and exit out the top through a Tenax trap (14 cm. length X 2.2 cm i.d.; 10 g of Tenax [Alltech Associates, Deerfield, IL]). The sampling was continued at room temperature (ca. 24°C) for 24 h at 3 L/min. The collected volatiles were eluted from the Tenax trap with 100 mL of freshly distilled diethyl ether containing ca. 0.001% Ethyl antioxidant 330 (1,3,5-trimethyl-2,4,6-tris(3,5-di-tert-butyl-4-hydroxybenzyl)benzene).

2. Simultaneous Distillation-Extraction (SDE). The washed fruit was cut and the stones were removed. The skin and pulp (1.02 kg) were blended with 1500 mL of water in a Waring blender for 15 s. 4-Nonanone (1 µl) was added as an internal standard and the mixture was blended for an additional 15 s. The mixture was subjected to SDE for 2 h with pentane-diethyl ether (1:1, v/v) using the SDE head described by Schultz et al. (6). The extract was dried over anhydrous sodium sulfate and concentrated with a Vigreux column to a final volume of 0.2-0.3 mL.

3. Vacuum SDE. The skin and pulp (3.18 kg, stones removed) were blended with 3500 mL of water in a Waring blender for 15 s. 4-Methylpent-2-yl acetate and 4-nonanone (final concentration - 0.8 ppm each relative to fruit weight) were added to the fruit slurry as internal standards. The slurry was blended for an additional 15 s and then added

to a 12 L round-bottomed flask. Eighty milliliters of antifoam solution was added to the flask. The antifoam solution was prepared by adding 10 mL of Hartwick antifoam 50 emulsion to 1000 mL of water and boiling until the volume was reduced to ca. 700 mL to remove volatiles. Vacuum SDE was performed (60 mmHg) with hexane over 2 h using the SDE head described by Schultz et al. (6). The extract was frozen (-20°C) to remove residual water and concentrated under reduced pressure with a Vigreux column to a final volume of 0.8-1.0 mL.

4. XAD-2 Column. Amberlite XAD-2 (20-60 mesh) was cleaned by successive extractions in a Soxhlet apparatus with pentane, ethyl acetate and methanol (each for 8 h). The slurry was filled into a glass column (2.5 cm i.d. X 53 cm, bed height - 33 cm).

5. Isolation of Glycosides with XAD-2 Column (7). Fruit pulp and skin (1 kg) were homogenized with NaCl (700 g) and water (1 L) in a Waring blender. The homogenate was centrifuged at 3200 rpm for 20 min. The resulting supernatant solution (1000 mL was adjusted to pH 5 with 3 N NaOH solution. The pH adjusted juice was diluted with 700 mL water and passed through filter paper. The filtrate was then loaded on the Amberlite XAD-2 column which had been just previously washed with 500 mL of water. After the filtrate had passed onto the resin the column was washed with 2200 mL of water followed by 600 mL of pentane. The glycosides were obtained by elution with 800 mL of ethyl acetate followed by 600 mL of methanol. The ethyl acetate and methanol eluates were each divided into halves; one half was subjected to SDE and the other half was used for enzymatic hydrolysis.

6. Acid Hydrolysis of Glycosides. Each solvent eluate (ethyl acetate and methanol) was concentrated to dryness on a rotary evaporator and treated separately. Each glycoside fraction was dissolved in 0.2 M phosphate buffer (200 mL, pH 3.0) and subjected to continuous liquid-liquid extraction (16 h) with diethyl ether to remove any volatiles. The aqueous glycoside solution was subjected to SDE for 2 h with pentane:ether (1:1, v/v) using the SDE head described by (6). The extracts were dried over anhydrous sodium sulfate and concentrated with a Vigreux column to a final volume of 0.1-0.2 mL.

7. Enzymatic Hydrolysis of Glycosides. Each solvent eluate (ethyl acetate and methanol) was taken to dryness on a rotary evaporator and treated separately. Each glycoside fraction was dissolved in 160 mL of 0.2 M phosphate buffer (pH 5.0) and continuously extracted (16 h) with diethyl ether to remove any volatiles. ß-Glucosidase (25 mg; Carl Roth KG, Karlsruhe, ca. 310 U/mg) was added and the mixture was stirred for 33 h at 36°C. The liberated

aglycons were removed by continuous liquid-liquid extraction (16 h) with diethylether. After drying and concentration the extract was ready for analysis. Pectinase (70 mg, Rohapect C, Rohm Tech., Inc.) was added to the previously enzyme (ß-glucosidase) hydrolyzed glycoside solution. The mixture was stirred for 50 h at 36°C. Released aglycons were removed by continuous liquid-liquid extraction (16 h) with diethyl ether. The ether extract was dried and concentrated in the manner previously described.

8. Isolation of Polar Constituents by Continuous Liquid-Liquid Extraction. Nectarine juice was prepared in the same way as for the isolation of glycosides. The pH adjusted juice (pH 5.0) was subjected to continuous liquid-liquid extraction with pentane for 22 h. The juice was then extracted with diethyl ether for 18 h to isolate the polar constituents. The ether extract was dried over anhydrous sodium sulfate and concentrated to 1 mL on a Vigreux column.

Capillary Gas Chromatography. A Hewlett-Packard 5890 gas chromatograph equipped with a flame ionization detector (FID) was used. A 60 m X 0.32 mm (i.d.) DB-1 fused silica capillary column (d_f - 0.25 μm, J&W Scientific, Folsom, CA) was employed. The injector and detector temperatures were 180 °C and 260°C, respectively. The oven temperature was programmed from 30 °C (4 min isothermal) to 240°C at 2 °C/min. The helium carrier gas linear velocity was 33-35 cm/s (30°C).

Capillary Gas Chromatography-Mass Spectrometry (GC/MS). Two different systems were used. The first system was a Finnigan MAT 4500 GC/MS/INCOS system (Finnigan MAT, San Jose, CA) equipped with the same type of column used in the GC analyses. The oven temperature was programmed from 50°C to 250°C at 2°C/min. For the hexane extract (obtained by vacuum SDE of blended fruit) the oven temperature was programmed from 30°C to 200°C at 2°C/min. The instrument was operated in the electron impact mode at an ionization voltage of 70 eV. The second system consisted of a HP 5890 gas chromatograph coupled to a HP 5970B quadrupole mass spectrometer (capillary direct interface). A 60 m X 0.25 mm (i.d.) DB-1 fused silica capillary column (d_f - 0.25 μm) was used. The oven temperature was programmed from 30°C (4 min isothermal) to 210°C at 2°C/min. Both systems utilized helium as the carrier gas.

Reference Compounds. Reference standards were obtained commercially, synthesized by established methods or received as gifts. (E)-2-hexenyl octanoate had the following mass spectrum: 226(M^+, 3), 197(4), 183(2), 155(5), 142(2), 128(9), 127(100), 125(6), 109(10), 100(5),

97(8), 83(28), 82(28), 67(45), 57(93), 55(71), 54(12), 43(24), 41(66). Mandelonitrile, prepared according to the method of Hirvi and Honkanen (8), had the following mass spectrum: 133(76, M^+), 132(50), 116(29), 115(43), 106(44), 105(100), 79(47), 78(37), 77(95), 51(68), 50(46), 39(19). Vomifoliol, prepared by $NaBH_4$ reduction of dehydrovomifoliol according to the method of Sefton et al. (9), had the following mass spectrum: 168(7), 166(2), 151(5), 150(11), 149(4), 135(10), 125(9), 124(100), 122(11), 111(7), 107(11), 95(7), 94(7), 91(8), 79(16), 77(10), 69(8), 55(11), 43(29). 7,8-Dihydrovomifoliol, prepared by the reduction of vomifoliol according to the method of Sefton et al. (9), had the following mass spectrum: 208(3), 193(2), 183(2), 170(29), 166(6), 153(29), 152(58), 125(26), 111(69), 110(100), 109(31), 107(28), 96(31), 82(24), 69(21), 68(23), 55(23), 45(22), 43(50). 3-Oxo-α-ionol had the following mass spectrum: 193(2), 175(1), 165(2), 153(2), 152(13), 135(7), 109(27), 108(100), 95(10), 91(13), 79(12), 77(11), 67(7), 65(6), 55(6), 45(14), 43(21). 3-Hydroxy-β-ionone had the following mass spectrum: 208(M^+, 5), 194(13), 193(100), 175(30), 149(10), 147(11), 133(11), 131(14), 121(13), 109(14), 105(22), 91(24), 79(14), 77(15), 55(9), 43(44). 3-Hydroxy-7,8-dihydro-β-ionol had the following mass spectrum: 212(M^+, 3), 194(4), 179(6), 162(5), 161(36), 153(7), 137(28), 136(33), 135(17), 123(16), 121(100), 119(76), 109(25), 107(29), 105(35), 95(31), 93(43), 91(29), 81(29), 79(27), 77(16), 69(23), 67(28), 55(28), 43(34), 41(39). 3-Oxo-7,8-dihydro-α-ionol, prepared by the reduction of 3-oxo-α-ionol (10 mg) with palladium black (5 mg) in 10 mL ethyl acetate (1 h reaction time, 3 psi H_2) had a mass spectrum which was in good agreement with literature data (48). 3-Oxo-retro-α-ionols, prepared according to the method of Herderich and Winterhalter (66), had mass spectra consistant with that previously reported (66). Isomer 1 and isomer 2 had Kovats' retention indices (DB-1) of 1663 and 1718, respectively. trans-Marmelo lactone had the following mass spectrum: 166(M^+, 28), 151(11), 138(15), 124(11), 123(40), 109(20), 107(12), 97(23), 96(27), 95(27), 93(42), 91(29), 81(34), 80(24), 79(27), 77(27), 69(39), 68(100), 67(43), 53(19), 42(36), 41(40). cis-Marmelo lactone had the following mass spectrum: 166(M^+, 35), 151(12), 138(16), 124(12), 123(43), 109(20), 107(13), 97(24), 96(29), 95(28), 93(49), 91(31), 81(34), 80(24), 79(28), 77(28), 69(39), 68(100), 67(43), 53(18), 42(34), 41(39). (Z)-2,6-Dimethylocta-2,7-diene-1,6-diol, prepared from linalool according to the method of Behr et al. (10), had the following mass spectrum: 137(8), 123(4), 121(5), 119(16), 110(9), 109(7), 107(5), 105(7), 97(5), 96(15), 95(13), 93(12), 91(7), 84(16), 83(9), 82(19), 81(15), 79(14), 71(57), 69(11), 68(19), 67(49), 57(11), 55(32), 53(14), 43(100), 41(40). (E)-2,6-Dimethylocta-2,7-diene-1,6-diol, prepared from linalool according to the method of Behr et al. (*10*), had the following mass spectrum: 152(1), 137(8), 123(4), 121(6), 119(13), 110(9), 109(8), 107(4), 105(5), 97(5), 96(12), 95(8), 94(8), 93(16), 91(9), 84(10), 83(8), 82(16), 81(13), 79(19), 71(59), 69(11), 68(20), 67(50), 57(10), 55(34), 53(16), 43(100), 41(37). The isomeric p-menth-1-en-9-als were prepared by subjecting (E)-2,6-dimethylocta-2,7-diene-1,6-diol to SDE for 2 h in 0.2 M phosphate buffer, pH 3.0. The aldehydes (isomers not separated) gave an ^{1}H NMR spectrum consistant with that previously reported (*49*). p-Menth-1-en-9-al (isomer 1) had the following mass spectrum: 95(17), 94(100), 93(11), 91(13), 81(12), 80(9), 79(93), 77(21), 68(13), 67(30), 55(15), 53(16), 41(17), 39(21). p-Menth-1-en-9-al (isomer 2) had the following mass spectrum: 95(16), 94(100), 93(11), 91(10), 81(11), 80(7), 79(93), 77(16), 68(12), 67(27), 65(9), 55(18), 53(18), 41(17), 39(20). Coniferyl alcohol had the following mass spectrum: 180(M^+, 67), 162(9), 152(10), 147(14), 138(11), 137(100), 131(18), 124(69), 119(32), 109(21), 107(12), 105(10), 103(27), 91(58), 89(12), 77(28), 65(27), 63(13), 55(16), 53(13), 51(20), 39(19). Dihydroconiferyl alcohol, prepared by the reduction of coniferyl alcohol (10 mg) with palladium black (5 mg) in 7 ml of absolute ethanol (2 h reaction time, 1 atm H_2), had the following mass spectrum: 182(M^+, 44), 164(4), 149(6), 139(6), 138(33), 137(100), 123(11), 122(18), 107(10), 94(14), 91(17), 77(18), 65(10), 51(10), 39(11).

Odor Threshold Determinations. These were determined (with reference standards purified by preparative gas chromatography) with a panel of 16-22 members using procedures described previously (*11*).

Results and Discussion

Volatile constituents of fresh, white-fleshed nectarines (experimental cultivar P 89-56) were isolated by headspace sampling of intact fruit, vacuum SDE of blended fruit and direct ether extraction of juice. Sample constituents were identified by comparison of the compound's Kovats index, I (*12*) and mass spectrum with that of a reference standard.

Headspace Sampling. Table I shows the constituents identified by dynamic headspace sampling of the intact fruit. These values should be considered as approximate since sample constituents coeluted with solvent peaks, and sample breakthrough during trapping was not determined.

Esters comprised a substantial portion of the volatiles, with large amounts of the following: (Z)-3-hexenyl acetate (15.99%), ethyl octanoate (10.72%), ethyl acetate (6.47%), hexyl acetate (2.61%), ethyl (Z)-4-

Table I. Headspace Constituents of Intact, Tree-Ripened Nectarine

Constituent	I^{DB-1} exp.	ref.	% area[a]
ethyl acetate	600	600	6.47
3-hydroxy-2-butanone	674	674	0.30
propyl acetate	694	695	0.19
toluene	752	748	0.36
2-methylpropyl acetate	763	764	1.11
hexanal	778	778	0.16
ethyl butanoate	788	789	0.19
(Z)-3-hexenol	843	843	0.29
ethylbenzene	847	844	0.22
m-xylene	856	852	0.59
hexanol	860	860	0.09
3-methylbutyl acetate	865	866	0.20
styrene	872	871	0.06
o-xylene	876	876	0.34
pentyl acetate	895	895	0.10
3-methylbut-2-enyl acetate	900	902	0.08
methyl hexanoate	909	910	0.06
benzaldehyde	924	926	0.08
propylbenzene	939	938	0.09
ethylmethylbenzene	947		0.29
1,3,5-trimethylbenzene[b]	955	953	0.13
1,2,4-trimethylbenzene	977	978	0.62
ethyl hexanoate	985	986	0.62
(Z)-3-hexenyl acetate	991	986	15.99
(E)-2-hexenyl acetate +	998	997	
hexyl acetate	998	997	2.61
γ-hexalactone	1000	1003	3.14
(monoterpene)[c]	1018		0.14
ethyl heptanoate	1083	1080	0.21
linalool	1085	1083	5.14
methyl (E)-4-octenoate[b]	1096	1094	0.23
methyl octanoate	1108	1107	1.33
naphthalene	1151	1157	0.26
(Z)-3-hexenyl butanoate	1170	1167	0.23
ethyl (E)-4-octenoate	1172	1172	0.43
ethyl octanoate	1183	1182	10.72
γ-octalactone	1205	1210	0.30
ethyl (E)-2-octenoate	1225	1223	0.10
δ-octalactone	1230	1234	0.11
2-methylnaphthalene	1261	1268	0.07
1-methylnaphthalene	1275	1283	0.06
theaspirane A[b]	1284	1284	1.71
methyl (Z)-4-decenoate	1290	1289	0.27
theaspirane B[b]	1298	1297	1.61
ethyl (Z)-4-decenoate	1363	1361	2.32
(E)-2-hexenyl hexanoate	1370	1368	0.21
ethyl decanoate	1380	1379	0.13

Table I. Headspace Constituents of Intact, Tree-Ripenend Nectarine (Continued)

Constituent	I^{DB-1} exp.	ref.	% area
6-pentyl-α-pyrone	1410	1415	0.71
dihydro-β-ionone[b]	1414	1414	2.99
γ-decalactone	1421	1423	6.86
δ-decalactone	1445	1448	2.18
epoxy-β-ionone[b]	1456	1456	0.10
β-ionone[b]	1459	1462	3.32
(Z)-3-hexenyl heptanoate	1461	1460	0.07
pentadecane	1501	1500	4.44
(Z)-3-hexenyl benzoate[b]	1539	1542	0.19
(Z)-3-hexenyl octanoate	1560	1559	1.72
(E)-2-hexenyl octanoate[b]	1566	1566	0.98
hexadecane	1600	1600	0.22
γ-dodecalactone	1631	1636	0.49
heptadecane	1701	1700	4.34
nonadecane	1900	1900	0.49

[a]Peak area percentage of total FID area excluding the solvent peaks (assuming all response factors of 1). [b]Identified for the first time in nectarine. [c]Tentative or partial identifications enclosed in parentheses.

decenoate (2.32%), (Z)-3-hexenyl octanoate (1.72%), methyl octanoate (1.33%) and 2-methylpropyl acetate (1.11%).

Notable was the presence of various C13 norisoprenoid compounds. These compounds were not previously found in yellow-fleshed nectarine (2) and may be responsible, in part, for differences in flavor characteristics. The isomeric theaspiranes have been described as possessing interesting sensory properties (13) though their flavor and fragrance value has been recently questioned (14). Preliminary studies in our lab (Takeoka and Buttery, unpublished data) indicate that these isomers are perhaps not as potent (odor thresholds 50-150 ppb) as had been previously believed (15).

Vacuum SDE. Nectarine volatiles identified in samples prepared by vacuum SDE are listed in Table II. Since the enzyme systems were not inhibited before disruption of the fruit tissues, the formation of secondary volatiles was reflected by the high levels of C6 lipid peroxidation products (16).

Many of the constituents identified were present in the headspace volatiles of intact fruit. Three additional lactones were detected. Jasmine lactone has been previously found in various nectarine cultivars (1). The tentatively identified (Z)-dodec-6-en-4-olide has been reported in butter (17), lamb (18), yeast (19) and in the male tarsal scent of blacktail deer (20). δ-Tetradecalactone has been found to occur in butter (17).

Two additional C13 norisoprenoids were found. 3,4-Didehydro-β-ionol has been identified in quince essential oil (21) and apricot (22). Its role as a flavor precursor will be discussed later. β-Damascenone is identified for the first time in nectarine though its presence in many other products is well established (13). Due to its low odor threshold of 0.002 ppb (23) this compound is probably an important flavor contributor in nectarine (500 odor units).

Eugenol was not recovered by this sampling method though it was found to exist in the free state as shown by direct ether extraction. There was also low recovery of other polar constituents such as lactones as has also been demonstrated in other studies (24).

Atmospheric SDE. While this technique exposes the sample to high temperatures and has a high probability of artifact formation it was nevertheless used for several reasons. First, it can simulate the composition of volatiles that would be formed as the result of processing (i.e. canning). Second, it is known that there is a poor recovery of certain polar compounds (i.e. lactones) with vacuum SDE. Third, knowledge of the volatiles formed during thermal treatment will give information on the type and amount of bound flavor constituents present.

Table III lists the nectarine volatiles identified by

Table II. Volatile Constituents of Nectarine: Vacuum
Steam Distilled Blended Fruit

Constituent	I^{DB-1} exp.	ref.	Approx. conc.[a] ug/kg
hexanal	780	778	226
(E)-2-hexenal	831	827	757
(Z)-3-hexenol	844	843	49
(E)-2-hexenol	859	856	760
hexanol	866	860	818
benzaldehyde	927	926	434
(Z)-3-hexenyl acetate	991	986	82
(E)-2-hexenyl acetate +	998	997	
hexyl acetate	998	997	97
phenylacetaldehyde[b]	1003	1002	6
nonanal[b]	1083	1082	3
linalool	1085	1083	10
methyl octanoate	1108	1107	2
nonanol	1158	1154	1
α-terpineol	1168	1170	1
(E)-2-hexenyl butanoate[b]	1177	1174	4
ethyl octanoate	1182	1180	2
γ-octalactone	1205	1210	7
(E)-2-decenal[b]	1235	1236	3
theaspirane A[b]	1284	1284	6
(E,E)-2,4-decadienal[b]	1285	1287	2
methyl (Z)-4-decenoate	1290	1289	1
theaspirane B[b]	1297	1297	7
ß-damascenone[b]	1360	1360	1
(Z)-3-hexenyl hexanoate	1362	1360	2
3,4-didehydro-ß-ionol[b]	1392	1397	2
(E)-2-hexenyl hexanoate	1370	1368	3
6-pentyl-α-pyrone	1412	1415	37
dihydro-ß-ionone[b]	1414	1414	24
γ-decalactone +	1427	1423	
geranylacetone[b]	1427	1422	376
(Z)-dec-7-en-5-olide	1440	1442	4
δ-decalactone	1447	1447	55
epoxy-ß-ionone[b]	1456	1456	1
ß-ionone[b]	1459	1462	16
(Z)-3-hexenyl benzoate[b]	1539	1542	3
(Z)-3-hexenyl octanoate	1560	1559	1
((Z)-dodec-6-en-4-olide)[b,c]	1604		5
γ-dodecalactone	1633	1635	49
δ-tetradecalactone[b]	1870	1870	1

[a]Only approximate concentrations since percent recoveries
and FID response factors were not determined for each
compound (assuming all response factors of 1). [b]Identi-
fied for the first time in nectarine. [c]Tentative or
partial identifications enclosed in parentheses.

Table III. Volatile Constituents of Nectarine: Atmospheric
Steam Distillation - Extraction

Constituent	I^{DB-1} exp.	ref.	Approx. conc.[a] ug/kg
propyl acetate	694	695	73
3-penten-2-one[b]	715	711	2
1,1-diethoxyethane[b]	725	716	16
(E)-2-pentenal[b]	728	723	45
(E)-2-pentenol	760	744	42
2-methylpropyl acetate	763	751	70
hexanal	783	772	7195
furfural[b]	800	800	889
(Z)-2-hexenal	821	817	492
(E)-2-hexenal	838	822	16170
(Z)-3-hexenol	846	834	162
(E)-2-hexenol	860	844	3655
hexanol	868	848	4284
benzaldehyde	927	926	923
(E)-2-heptenal[b]	929	927	19
6-methyl-5-hepten-2-one	967	966	13
2-pentylfuran	981	977	61
(Z)-3-hexenyl acetate	990	986	37
(E)-2-hexenyl acetate +	998	997	
hexyl acetate	998	997	227
γ-hexalactone	998	1003	152
benzyl alcohol +	1002	1004	
phenylacetaldehyde[b]	1002	1004	113
(E)-2-octenal[b]	1032	1030	–
4-nonanone(IS)			
(E)-linalool oxide furanoid	1059	1056	17
(Z)-linalool oxide furanoid[b]	1071	1070	7
nonanal[b]	1083	1082	146
hotrienol	1085	1085	21
3-nonen-2-one[b]	1115	1114	133
(E,Z)-2,6-nonadienal[b]	1124	1124	8
(E)-2-nonenal[b]	1134	1133	23
3a,4,5,7a-tetrahydro-4,4,7a-trimethyl-1-methylene-1H-indene[b]	1178	1178	12
ethyl octanoate	1183	1180	10
p-menth-1-en-9-al (isomer 1)[b]	1185	1185	44
p-menth-1-en-9-al (isomer 2)[b]	1187	1187	35
2,2,6,7-tetramethylbicyclo[4.3.0]nona-4,7,9-(1)-triene[b]	1196	1195	585
γ-octalactone	1205	1210	32
δ-octalactone	1232	1234	tr

Table III. Volatile Constituents of Nectarine: Atmospheric
Steam Distillation - Extraction (Continued)

Constituent	I^{DB-1} exp.	ref.	Approx. conc. ug/kg
(E)-2-decenal[b] +	1236	1236	
MW-174	1236		37
MW-174	1251		25
nonanoic acid[b]	1261	1260	25
vitispiranes[b]	1263	1263	43
theaspirane A[b]	1284	1284	13
(E,E)-2,4-decadienal[b]	1286	1287	37
MW-174	1290		157
MW-174	1293		19
(5-undecen-4-one)[b,c]	1295		17
MW-174	1305		28
γ-nonalactone	1311	1315	14
eugenol[b]	1324	1326	77
1,1,6-trimethyl-1,2-dihydronaphthalene[b]	1328	1329	28
MW-174	1332		28
1,1,6-trimethyl-1,2,3,4-tetrahydronaphthalene[b]	1334	1334	149
(megastigma-4,6,8-triene)[b,c]	1346		12
ß-damascenone[b]	1358	1360	29
trans-marmelo lactone[b]	1362	1362	120
cis-marmelo lactone[b] + ?	1367	1368	160
(E)-2-hexenyl hexanoate[b]	1370	1368	10
3,4-didehydro-ß-ionol[b]	1393	1397	337
6-pentyl-α-pyrone	1411	1415	155
dihydro-ß-ionone[b]	1413	1414	37
γ-decalactone	1423	1422	1207
geranylacetone	1428	1428	53
(Z)-dec-7-en-5-olide [jasmin lactone]	1442	1442	tr
δ-decalactone	1446	1447	240
8,9-dehydrotheaspirone	1456	1455	65
ß-ionone[b]	1459	1462	16
(4,4,7-trimethyl-3,4-dihydro-2(1H)-naphthalenone)[b,c]	1507		20
(phthalate)[c]	1547		14
dodecanoic acid[b] + ?	1549	1547	18
ethyl dodecanoate[b]	1579	1578	12
(megastigma-4,6,8-triene-3-one)[b,c]	1585		10
((Z)-dodec-6-en-4-olide)[b]	1604		26
γ-dodecalactone	1632	1635	232
tetradecanoic acid[b]	1742	1740	6
ethyl tetradecanoate[b]	1779	1778	6

Continued on next page

Table III. Volatile Constituents of Nectarine: Atmospheric
Steam Distillation - Extraction (Continued)

Constituent	I^{DB-1} exp.	ref.	Approx. conc. ug/kg
δ-tetradecalactone[b]	1870	1871	27
hexadecanoic acid[b]	1953	1940	410
ethyl hexadecanoate[b]	1979	1978	69
n-heneicosane[b] [C21]	2100	2100	26
ethyl oleate[b]	2148	2146	26
ethyl octadecanoate[b]	2179	2176	46
n-tricosane[b] [C23]	2301	2300	304
n-tetracosane[b] [C24]	2400	2400	22

[a]Only approximate concentrations since percent recoveries and FID response factors were not determined for each compound (assume all response factors of 1). tr represents concentration less than 1 ppb. [b]Identified for the first time in nectarine. [c]Tentative or partial identifications enclosed in parentheses.

atmospheric SDE of blended fruit. There were drastically higher levels of the C6 lipid peroxidation products especially hexanal and (E)-2-hexenal. There were increased amounts of other aldehydes such as (E)-2-heptenal and nonanal. The sugar degradation product, furfural, was also observed. The ketones, 6-methyl-5-hepten-2-one and 3-nonen-2-one were present. There were also higher levels of the lactones found. The level of γ-decalactone was more than 3 times higher and the level of δ-decalactone was more than 4 times higher than that found by vacuum SDE. This not only reflects the better recovery of free lactones but also the release of glycosidically bound forms.

There was a dramatic increase in the levels of 3,4-didehydro-ß-ionol and the potent ß-damascenone. Notable was the presence of the isomeric p-menth-1-en-9-als, 2,2,6,7-tetramethylbicyclo[4.3.0]nona-4,7,9(1)-triene, vitispirane, 1,1,6-trimethyl-1,2-dihydronaphthalene, 1,1,6-trimethyl-1,2,3,4-tetramethylnaphthalene and marmelo lactones. These constituents were derived from the acid hydrolysis of glycosides as will be demonstrated later. Trace amounts of 3a,4,5,7a-tetrahydro-4,4,7a-trimethyl-1-methylene-1H-indene which was recently identified in quince brandy (59) were also found.

Megastigma-4,6,8-triene-3-one was probably formed from the acid catalyzed degradation of 3-oxo-α-ionol (25). The latter ionol was found to be glycosidically bound in P 89-56. While no compounds eluting above C19 were observed in the sample prepared by vacuum SDE various hydrocarbons, free acids and esters with Kovats indices higher than 1900 were identified in the SDE sample. There was evidence of free acids as shown by front-tailing peaks present. Most of these were not quantitated since they overlapped with other peaks.

Odor Units Values from Vacuum and Atmospheric SDE. The relative importance of various compounds to the overall aroma was determined by calculating the number of odor units. The odor unit has been defined as the compound concentration divided by its odor threshold (26). The results of these calculations are listed in Table IV. Examining the sample prepared by vacuum SDE reveals that ß-ionone and ß-damascenone are prominent odor contributors. While the role of lactones to nectarine and peach flavor has been previously discussed (1,27), this study shows the importance of these C13 norisoprenoids and their possible contribution to the unique flowery character of this cultivar. With the sample prepared by atmospheric SDE ß-damascenone became the dominant contributor followed by ß-ionone, hexanal, (E)-2-hexenal, (E)-2-nonenal, 1,1,6-trimethyl-1,2,3,4-tetrahydronaphthalene, nonanal, γ-decalactone, hexyl acetate and γ-dodecalactone. ß-Damascenone and 1,1,6-trimethyl-1,2,3,4-tetrahydronaphthalene are glycoside hydrolysis products as will be discussed later.

Table IV. Odor Thresholds in Water and Odor Units of Some Nectarine Constituents

Constituent	Odor Threshold (ppb)	Odor Units (U_o)[a] Vacuum SDE	SDE
ß-ionone	0.007[b]	2286	2286
ß-damascenone	0.002[c]	500	14500
hexanal	5[b]	45	1439
(E)-2-hexenal	17[b]	45	951
γ-decalactone	10	38	121
(E,E)-2,4-decadienal	0.07[b]	29	–
(E)-2-decenal	0.4	7.5	–
γ-dodecalactone	7[d]	7	33
nonanal	1	3	146
linalool	6[e]	1.7	–
phenylacetaldehyde	4[b]	1.5	–
benzaldehyde	350[b]	1.2	2.6
γ-octalactone	8	0.9	4.6
(Z)-3-hexenol	70[b]	0.7	–
δ-decalactone	100[d]	0.6	2.4
hexanol	2500	0.3	1.7
6-pentyl-α-pyrone	200	0.2	1.0
nonanol	50	0.02	–
epoxy-ß-ionone	100	0.01	–
methyl octanoate	200	0.01	–
(Z)-dec-7-en-5-olide	1000	0.004	–
α-terpineol	330	0.003	–
(E)-2-nonenal	0.08[b]	–	288
1,1,6-trimethyl-1,2-3,4-tetrahydronaphthalene	2[f]	–	149
hexyl acetate	2	–	113
1,1,6-trimethyl-1,2-dihydronaphthalene	2.5[f]	–	11
2-pentylfuran	6	–	10
(E)-2-hexenol	400	–	9
2-methylpropyl acetate	65	–	1
geranylacetone	60[b]	–	0.88
6-methyl-5-hepten-2-one	50[b]	–	0.26
3-nonen-2-one	800	–	0.17
γ-hexalactone	1600[d]	–	0.10
δ-tetradecalactone	135	–	0.04
furfural	23000	–	0.04

[a]Uo = compound concentration divided by its odor threshold. [b](28). [c](23). [d](1). [e](29). [f](30).

Free Polar Constituents Identified in the Ether Extract. Constituents listed in Table V are those not previously found by vacuum SDE. 2-Methylbutanoic acid, 3-methylbutanoic acid, benzoic acid and eugenol also existed in the bound form. 8,9-Dehydrotheaspirone, which is reported to have a strong flowery-woody odor, has been previously identified in *Nicotiana tabacum* (31). It has also been found to be enzymatically released from grape juice glycosides (32). Dehydrovomifoliol has been characterized in kidney bean roots (33), tea (34), grapes (25) and quince (35). There were additional C13 norisoprenoids (as indicated by their MS) present in the extract which could not be identified.

Table V. Compounds Identified by Direct Ether Extraction Not Found by Vacuum SDE

Compound	I^{DB-1} exp.	ref.	%area[a]
ethyl acetate	604	600	7.36
2-pentanone	666	657	0.15
3-hydroxy-2-butanone	681	674	5.78
2-methylpropanoic acid	778	778	0.28
butanoic acid	805	805	0.02
3-methylbutanoic acid	866	842	1.21
2-methylbutanoic acid	883	865	1.77
γ-hexalactone	1000	1003	0.96
benzoic acid	1157	1145	0.78
2-hydroxybenzoic acid	1285	1282	0.03
eugenol	1325	1327	0.19
8,9-dehydrotheaspirone	1460	1455	9.46
dehydrovomifoliol	1745	1723	19.46
tetradecanoic acid	1760	1740	0.04
hexadecanoic acid	1954	1939	5.04
ethyl hexadecanoate	1978	1978	0.07

[a]Peak area percentage of total FID area excluding the solvent peaks (assuming all response factors of 1).

Products of Enzymatic Hydrolysis (Table VI). Moderate amounts of the monoterpene diols were found. 2,6-Dimethyloct-7-ene-1,6-diol was previously identified in both the free and bound form in grapes from different cultivars (36). The bound form of this diol was recently found in Riesling wine (32). Bound forms of (E) and (Z)-2,6-dimethylocta-2,7-dien-1,6-diol have been found and characterized in a number of products. The ß-D-glucoside of both isomers has been identified in *Betula alba* leaves

Table VI. Compounds Identified in Nectarine Glycoside
Fraction (ethyl acetate eluate) After Enzymatic
Hydrolysis (ß-glucosidase)

Compound	I^{DB-1} exp.	ref.	%area
3-methylbutanol	725	725	0.08
3-methylbutanoic acid	860	842	-
2-methylbutanoic acid	876	850	2.34
benzaldehyde	929	926	39.24
benzyl alcohol	1003	1004	0.49
2-phenylethanol	1081	1081	0.06
3-nonen-2-one	1115	1114	0.65
benzoic acid	1150	1143	0.61
(ethylbenzaldehyde)[b]	1228		0.09
mandelonitrile	1269	1268	0.36
theaspirane A	1284	1284	0.03
theaspirane B	1298	1297	0.05
2,6-dimethyloct-7-ene-1,6-diol	1308	1307	1.35
(Z)-2,6-dimethylocta-2,7-diene-1,6-diol	1312	1309	2.17
eugenol	1324	1327	0.84
(E)-2,6-dimethylocta-2,7-diene-1,6-diol	1333	1331	5.04
vanillin	1345	1348	0.34
2-aminobenzoic acid	1358	1359	0.26
(3,4-dimethoxyphenol)[b]	1377		0.04
methyl 4-hydroxybenzoate	1419	1418	0.08
acetovanillone	1438	1437	0.20
δ-decalactone	1446	1447	0.38
dihydroconiferyl alcohol	1597	1596	0.31
3-oxo-α-ionol	1602	1600	0.58
3-hydroxy-7,8-dihydro-ß-ionol	1626	1622	8.75
3-hydroxy-ß-ionone	1647	1646	0.38
3-oxo-7,8-dihydro-α-ionol	1657	1655	0.66
coniferyl alcohol	1682	1680	1.41
3-oxo-retro-α-ionol(isomer 2)	1720	1718	0.28
vomifoliol	1736	1734	0.91
(3-hydroxy-1-(4-hydroxy-3-methoxy)-1-propanone)[b]	1748		0.50
7,8-dihydrovomifoliol	1793	1792	0.41

[a]Peak area percentage of total FID area excluding the
solvent peaks (assuming all response factors of 1).
[b]Tentative or partial identifications enclosed in
parentheses.

and *Chaenomeles japonica* leaves (*37*). In *Vitis vinifera* grapes, both the ß-D-glucopyranoside and the 6-O-α-arabinofuranosyl-ß-D-glucopyranoside of the (E) isomer have been found (*38*). The ß-D-glucoside of the (E) isomer has also been characterized in sour cherry (*39*). Bound forms of both isomers have been reported in papaya (*40*) while raspberry (*41*), apricot, peach and yellow plum (*4*) contain only conjugates of the (E) isomer.

Acetovanillone, coniferyl alcohol and dihydroconiferyl alcohol are known as glycosidically bound grape flavor constituents (*3,42*) while the conjugate of the latter compound has also been found in raspberry (*41*). The 3-O-ß-D-glucopyranoside of 3-hydroxy-1-(4'-hydroxy-3'-methoxyphenyl)-1-propanone has been isolated and identified in *Pinus sylvestris* (*43*).

Benzaldehyde, the major constituent liberated from the glycoside fraction (39.24%), has been previously reported to occur in the bound form in various *Prunus* species (*4*). Surprisingly, this compound occurred as only a minor constituent (0.31%) in the acid hydrolysate of the glycoside fraction (Table VII).

The isomeric theaspiranes, found in both the headspace and vacuum SDE samplings, have also been found to be enzymatically released from raspberry glycosides (*41*). These isomers may arise as artefacts from 4-hydroxy-7,8-dihydro-ß-ionol. The latter diol readily dehydrates under mild conditions in acidic media to yield the isomeric theaspiranes (*44*). Conjugated forms of 3-hydroxy-7,8-dihydro-ß-ionol, 3-hydroxy-ß-ionone and vomifoliol have been found in various *Prunus* species such as apricot, peach and yellow plum (*4*). The ß-D-glucoside of vomifoliol, known as roseoside, was isolated and characterized in *Vinca rosea* (*45*) and later found in *Betula alba*, *Cydonia oblonga* (*46*) and *Vitis vinifera* grapes (*42*). The 1-O-ß-D-xylopyranosyl 6-O-ß-D-glucopyranoside of vomifoliol was characterized in apples (cv. Jonathan) (*47*). Bound forms of 3-oxo-retro-α-ionol have been reported in grapes (*9*) and passion fruit (*48*). Conjugated 7,8-dihydrovomifoliol has been identified in grapes (*9*) and *Pinus sylvestris* needles (*43*). Dehydrovomifoliol, which was identified as a major constituent in the free state (ppm level, ether extract), was not found as a bound constituent.

<u>Products of Acid Hydrolysis</u> (Table VII). 3-Nonen-2-one was found in both the enzymatic and acid hydrolysates. This compound had been previously found in peaches prepared by atmospheric SDE (*27*) while we could not detect it in the free state.

The diastereomers of p-menth-1-en-9-al were also identified. This aldehyde was previously found in Bulgarian rose oil where it made a contribution to the odor (*49*). It has also been found in the essential oil of cotton buds (*50*) and Buchu leaf oil (*51*). It has been

Table VII. Compounds Identified in Nectarine Glycoside
Fraction (ethyl acetate eluate) After SDE Acid Hydrolysis
(pH 3.0)

Compound	I^{DB-1} exp.	ref.	%area[a]
3-methylbutanoic acid + ?	861	842	4.32
2-methylbutanoic acid	876	850	5.46
benzaldehyde	925	926	0.31
3-nonen-2-one	1115	1114	3.05
p-menth-1-en-9-al(isomer 1)	1185	1185	1.75
p-menth-1-en-9-al(isomer 2)	1187	1187	1.50
2,2,6,7-tetramethylbicyclo [4.3.0]nona-4,7,9(1)-triene	1195	1195	4.17
vitispiranes	1263	1263	1.14
MW-174	1290		1.41
eugenol	1326	1326	4.96
1,1,6-trimethyl-1,2-dihydro-naphthalene	1329	1329	0.49
1,1,6-trimethyl-1,2,3,4-tetrahydronaphthalene	1334	1334	1.44
ß-damascenone	1359	1360	0.47
trans marmelo lactone	1362	1362	4.39
cis marmelo lactone	1368	1368	4.66
(2,2,6,7-tetramethylbicyclo [4.3.0]nona-4,9(1)-dien-8-ol)[b]	1380		1.13
(2,2,6,7-tetramethylbicyclo [4.3.0]nona-4,9(1)-dien-8-ol isomer)[b]	1385		0.51
3,4-didehydro-ß-ionol	1393	1397	2.90
(4,5,6,7-tetrahydro-3,3a,7,7-tetramethyl-5(3aH)-indenol)[b]	1397		2.24
(4,5,6,7-tetrahydro-3,3a,7,7-tetramethyl-5(3aH)-indenol isomer)[b]			
ß-ionone	1459	1462	1.07

[a]Peak area percentage of total FID area excluding the
solvent peaks (assuming all response factors of 1).
[b]Tentative or partial identifications enclosed in
parentheses.

found in sour cherry, specifically when the glycoside fraction was subjected to SDE at pH 1 (5). The discovery of this aldehyde is not surprising since it has been shown that acid hydrolysis of the diol, (E)-2,6-dimethylocta-2,7-dien-1,6-diol gives (E,E)-2,6-dimethylocta-2,6-dien-1,8-diol, 3,9-epoxy-p-menth-1-ene and the isomeric p-menth-1-en-9-als as products (38).

Marmelo lactones were first identified in quince (*Cydonia oblonga* Mill.) essential oil by Tsuneya et al. (52) who reported them as key contributors to the characteristic odor. The lactones are major constituents of quince essential oil comprising 10.34% (trans) and 13.03% (cis) of the total volatiles (53). The lactones are released from the glycoside fraction by either enzymatic or acid-catalyzed hydrolysis of quince (35) and peach cv. Hakuto (Shiota, H., Shiono Koryo Kaisha Ltd., personal communication, 1989) and cv. Redhaven (4). ß-D-Glucopyranoside of 2,7-dimethyl-8-hydroxy-4(E), 6(E)-octadienoic acid was recently isolated and characterized in quince juice as the natural precusor of these important lactones (54).

The hydrocarbon, 1,1,6-trimethyl-1,2-dihydronaphthalene (TDN) has been previously found in strawberry fruit and foliage (55), passion fruit (56) and tomato paste (30). It has an odor threshold of 2.5 ppb in water (30) and has odor properties described as "hydrocarbon" or "kerosene-like" (57). The acid hydrolysis of the glycoside fraction from grape juice leads to formation of TDN (58). Though different compounds have been proposed as precursors of TDN, recent studies by Winterhalter (48) indicate that the bound form of 3-hydroxy-1,1,6-trimethyl-1,2,3,4-tetrahydronaphthalene is the most likely precursor of TDN in purple passion fruit. Winterhalter (60) has also identified glycosidically bound forms of 2,6,10,10-tetramethyl-1-oxaspiro-[4.5]dec-6-ene-2,8-diol in Riesling wine and discussed its role as a natural precursor of TDN.

2,2,6,7-Tetramethylbicyclo[4.3.0]nona-4,7,9(1)-triene was first identified in quince essential oil (21). Its immediate precursor appears to be 3,4-didehydro-ß-ionol as refluxing this compound in acidic solution produces an 80% yield of the bicyclic hydrocarbon (21). Glycosidically bound 3-hydroxy-ß-ionol has been postulated as a likely precursor of 3,4-didehydro-ß-ionol in quince (35). In fact, the ß-D-gentiobioside [ß-D-glucopyranosyl(1→6)-ß-D-glucopyranoside] of 3-hydroxy-ß-ionol was recently isolated and characterized in quince (61). However, it appears that there must be another precursor in nectarine since the corresponding ionol was not detected after enzymatic hydrolysis of the glycosidic fraction.

The isomeric vitispiranes were first identified in grape juice, wine and distilled grape spirits (62). It has been shown that these isomers are important aroma constituents of vanilla and that the cis isomer has a fresher and more intense odor than the trans isomer (63). Studies by (64,65) indicate that a glycoside of 8-

hydroxytheaspiranes   (3-hydroxytheaspiranes   using
megastigmane numbering) is the most likely precursor of
vitispiranes in grapes.

Acknowledgment

We dedicate this chapter to Dr. Walter Jennings on the occasion of his 70th
birthday. We thank Haruyasu Shiota for supplying the sample of marmelo
lactones.

Literature Cited

1.   Engel, K.-H.; Flath, R.A.; Buttery, R.G.; Mon,
     T.R.; Ramming, D.W.; Teranishi, R. *J. Agric. Food
     Chem.* **1988**, *36*, 549-553.
2.   Takeoka, G.R.; Flath, R.A.; Guentert, M.; Jennings,
     W. *J. Agric. Food Chem.* **1988**, *36*, 553-560.
3.   Williams, P.J.; Sefton, M.A.; Wilson, B. In *Flavor
     Chemistry: Trends and Developments*; Teranishi, R.;
     Buttery, R.G.; Shahidi, F., Eds.; ACS
     Symposium Series 388; American Chemical Society:
     Washington, D.C. 1989, pp 35-48.
4.   Krammer, G.; Winterhalter, P.; Schwab, M.;
     Schreier, P. *J. Agric. Food Chem.* **1991**, *39*, 778-
     781.
5.   Schwab, W.; Schreier, P. *Z. Lebensm. Unters.-
     Forsch.* **1990**, *190*, 228-231.
6.   Schultz, T.H.; Flath, R.A.; Mon, T.R.; Eggling,
     S.B.; Teranishi, R. *J. Agric. Food Chem.* **1977**, *25*,
     446-449.
7.   Gunata, Y.Z.; Bayonove, C.L.; Baumes, R.L.;
     Cordonnier, R.E. *J. Chromatogr.* **1985**, *331*, 83-90.
8.   Hirvi, T.; Honkanen, E. Analysis of Volatile
     Constituents of Black Chokeberry (*Aronia
     melanocarpa* Ell.). *J. Sci. Food Agric.* **1985**, *36*,
     808-810.
9.   Sefton, M.A.; Skouroumounis, G.K.; Massy-Westropp,
     R.A.; Williams, P.J. *Aust. J. Chem.* **1989**, *42*, 2071-
     2084.
10. Behr, D., Wahlberg, I., Nishida, T., Enzell, C.R.
     Acta Chem. Scand. **1978**, B *32*, 228-234.
11. Guadagni, D.G.; Buttery, R.G. *J. Food Sci.* **1978**, *43*,
     1346-1347.
12. Kovats, E. sz. *Helv. Chim. Acta* **1958**, *41*, 1915-
     1932.
13. Ohloff, G. In *Progress in the Chemistry of Organic
     Natural Products*. Vol. 35, Herz, W.; Grisebach, H.;
     Kirby, G.W., Eds., Springer Verlag, New York, 1978,
     pp. 431-527.
14. Mookherjee, B.D.; Wilson, R.A. *Perf. & Flav.* **1990**,
     15, 27-33, 36-38, 40, 43-46, 48-49.
15. Naegeli, P. Ger. Offen. 2 610 238, 1976; Chem.
     Abstr. 1977, 86, 95876a.
16. Grosch, W. In *Food Flavours* Part A. Morton, I.D.;

MacLeod, A.J., Eds.; Elsevier, Amsterdam, 1982, pp. 325.

17. Boldingh, J.; Taylor, R.J. *Nature* **1962**, *194*, 909-913.
18. Park, R.J.; Murray, K.E.; Stanley, G. *Chem. Ind.* (London) **1974**, 380-382.
19. Takahara, S.; Fujiwara, K.; Mizutani, J. *Agric. Biol. Chem.* **1973**, *37*, 2855-2861.
20. Brownlee, R.G.; Silverstein, R.M.; Müller-Schwartze, D.; Singer, A.G. *Nature* **1969**, *221*, 284-285.
21. Ishihara, M.; Tsuneya, T.; Shiota, H.; Shiga, M.; Nakatsu, K. *J. Org Chem.* **1986**, *51*, 491-495.
22. Takeoka, G.R.; Flath, R.A.; Mon, T.R.; Teranishi, R.; Guentert, M. *J. Agric. Food Chem.* **1990**, *38*, 471-477.
23. Buttery, R.G.; Teranishi, R.; Ling, L.C. *Chem. Ind.* (London) **1988**, 238.
24. Horvat, R.J.; Chapman, G.W., Jr.; Robertson, J.A.; Meredith, F.I.; Scorza, R.; Callahan, A.M.; Morgens, P. *J. Agric. Food Chem.* **1990**, *38*, 234-237.
25. Strauss, C.R.; Wilson, B.; Williams, P.J. *Phytochemistry*, **1990**, *26*, 1995-1997.
26. Guadagni, D.G.; Buttery, R.G.; Harris, J. *J. Sci. Food Agric.* **1966**, *17*, 142-144.
27. Spencer, M.D.; Pangborn, R.M.; Jennings, W.G. *J. Agric. Food Chem.* **1978**, *26*, 725-732.
28. Buttery, R.G.; Seifert, R.M.; Guadagni, D.G.; Ling, L.C. *J. Agric. Food Chem.* **1971**, *19*, 524-529.
29. Buttery, R.G.; Seifert, R.M.; Guadagni, D.G.; Ling, L.C. *J. Agric. Food Chem.* **1969**, *17*, 1322-1327.
30. Buttery, R.G.; Teranishi, R.; Flath, R.A.; Ling, L.C. *J. Agric. Food Chem.* **1990**, *38*, 792-795.
31. Fujimori, T.; Takagi, Y.; Kato, K. *Agric. Biol. Chem.* **1981**, *45*, 2925-2926.
32. Winterhalter, P.; Sefton, M.A.; Williams, P.J. *J. Agric. Food Chem.* **1990**, *38*, 1041-1048.
33. Takasugi, M.; Anetai, M.; Katsui, N.; Masamune, T. *Chem. Lett.* **1973**, 245-248.
34. Etoh, H.; Ina, K.; Iguchi, M. *Agric. Biol. Chem.* **1980**, *44*, 2999-3000.
35. Winterhalter, P.; Schreier, P. *J. Agric. Food Chem.* **1988**, *36*, 1251-1256.
36. Versini, G.; Dalla Serra, A.; Dell'Eva, M.; Scienza, A.; Rapp, A. In *Bioflavour 87*; Schreier, P., Ed; de Gruyter: Berlin, New York, 1988; pp 161-170.
37. Tschesche, R.; Ciper, F.; Breitmaier, E. *Chem. Ber.* **1977**, *110*, 3111-3117.
38. Strauss, C.R.; Wilson, B.; Williams, P.J. *J. Agric. Food Chem.* **1988**, *36*, 569-573.
39. Schwab, W.; W.; Scheller, G.; Schreier, P. *Phytochemistry* **1990**, *29*, 607-612.
40. Schwab, W.; Mahr, C.; Schreier, P. *J. Agric. Food Chem.* **1989**, *37*, 1009-1012.
41. Pabst, A.; Barron, D.; Etiévant, P.; Schreier, P. *J Agric. Food Chem.* **1991**, *39*, 173-175.

42. Strauss, C.R.; Gooley, P.R.; Wilson, B.; Williams, P.J. *J. Agric. Food Chem.* **1987**, *35*, 519-524.
43. Andersson, R.; Lundgren, L.L. *Phytochemistry* **1988**, *27*, 559-562.
44. Winterhalter, P; Schreier, P. *J. Agric. Food Chem.* 1988, 36, 560-562.
45. Bhakuni, D.S.; Joshi, P.P.; Uprety, H.; Kapil, R.S. *Phytochemistry*, **1974**, *13*, 2541-2543.
46. Tschesche, R.; Ciper, F.; Harz, A. *Phytochemistry*, **1976**, *15*, 1990-1991.
47. Schwab, W.; Schreier, P. *Phytochemistry* **1990**, *29*, 161-164.
48. Winterhalter, P. *J. Agric. Food Chem.* **1990**, *38*, 452-455.
49. Ohloff, G.; Giersch, W.; Schulte-Elte, K.H.; sz. Kovats, E. *Helv. Chim. Acta* **1969**, *52*, 1531-1536.
50. Hedin, P.A.; Thompson, A.C.; Gueldner, R.C. *Phytochemistry* **1975**, *14*, 2087-2088.
51. Kaiser, R.; Lamparsky, D.; Schudel, P. *J. Agric. Food Chem.* **1975**, *23*, 943-950.
52. Tsuneya, T.; Ishihara, M.; Shiota, H.; Shiga, M. *Agric. Biol. Chem.* **1980**, *44*, 957-958.
53. Tsuneya, T.; Ishihara, M.; Shiota, H.; Shiga, M. *Agric. Biol. Chem.* **1983**, *47*, 2495-2502.
54. Winterhalter, P; Lutz, A.; Schreier, P. *Tetrahedron Lett.* **1991**, *32*, 3669-3670.
55. Stoltz, L.P.; Kemp, T.R.; Smith, W.O., Jr.; Smith, W.T., Jr.; Chaplin, C.E. *Phytochemistry* **1970**, *9*, 1157-1158.
56. Murray, K.E.; Shipton, J.; Whitfield, F.B. *Aust. J. Chem.* **1972**, *25*, 1921-1933.
57. Simpson, R.F. *Chem. Ind.* (London) **1978**, 37.
58. Williams, P.J.; Strauss, C.R.; Wilson, B.; Massy-Westropp, R.A. *J. Chromatogr.* **1982**, *235*, 471-480.
59. Näf, R.; Velluz, A. *J. Ess. Oil Res.* **1991**, *3*, 165-172.
60. Winterhalter, P. *J. Agric. Food Chem.* **1991**, *39*, 1825-1829.
61. Winterhalter, P.; Harmsen, S.; Trani, F. *Phytochemistry* **1991**, *30*, 3021-3025.
62. Simpson, R.F.; Strauss, C.R.; Williams, P.J. *Chem. Ind.* (London) **1977**, 663-664.
63. Schulte-Elte, K.-H.; Gautschi, F.; Renold, W.; Hauser, A.; Fankhauser, P.; Limacher, J.; Ohloff, G *Helv. Chim. Acta* **1978**, *61*, 1125-1133.
64. Strauss, C.R.; Williams, P.J.; Wilson, B.; Dimitriadis, E. In *Flavour Research of Alcoholic Beverages*; Nykänen, L.; Lehtonen, P., Eds.; Foundation for Biotechnological and Industrial Fermentation Research: Helsinki, 1984; pp 51-60.
65. Strauss, C.R.; Dimitriadis, E.; Wilson, B.; Williams, P.J. *J. Agric. Food Chem.* **1986**, *34*, 145-149.
66. Herderich, M.; Winterhalter, P. *J. Agric. Food Chem.* **1991**, *39*, 1270-1274.

RECEIVED December 18, 1991

THERMAL GENERATION

Chapter 11

Thermally Degraded Thiamin

A Potent Source of Interesting Flavor Compounds

Matthias Güntert, J. Brüning, R. Emberger, R. Hopp, M. Köpsel,
H. Surburg, and P. Werkhoff

Research Department, Haarmann and Reimer GmbH, Postfach 1253,
D–3450 Holzminden, Germany

Aqueous solutions of thiamin hydrochloride with different
concentrations and differently adjusted pH values were heated in an
autoclave for various times. The resulting flavor compounds were
obtained by applying the simultaneous distillation/extraction method
according to Likens-Nickerson. The flavor concentrate was preseparated by medium-pressure liquid chromatography on silica gel using
a pentane-diethyl ether gradient. The different fractions were subsequently analyzed by capillary gas chromatography (HRGC) and
capillary gas chromatography-mass spectrometry (HRGC/MS).
Various unknown compounds were isolated by preparative capillary
gas chromatography from the very complex mixtures in microgram-quantities to elucidate their structures by IR, NMR, and mass
spectrometry, and to check their olfactory properties.
The identified compounds were used to explain the various
degradation pathways of thermally treated thiamin. Their occurrence,
formation, sensory impression, and spectroscopic data are discussed.

Thiamin (vitamin B_1) belongs to the precursors which contribute to the pleasant
aroma of cooked and roasted meat. Its thermal degradation leads to numerous,
primarily S-containing flavor compounds. The literature about the degradation of this
interesting flavor precursor was exhaustively reviewed by us already in [1]. Remarkably, recent publications about the systematic investigation with modern analytical
chromatographic and spectroscopic equipment were not found. This was the reason to
start with the thiamin project. The authors are carrying out continuing studies to obtain a better understanding of the chemistry of the thermal degradation of thiamin and
the olfactory properties of the generated compounds. Moreover, during recent years
we investigated various thermally generated aromas in which thiamin played a more
or less important role for the overall odor impression [1-6].
This study shows primarily the different pathways of the thermal degradation of
thiamin at different pH values. Additionally, the effect of reaction time on the formation of degradation products is shown. An external standard method was used to
quantify the reaction mixtures. This makes it possible to compare not only
qualitatively but also quantitatively the generated volatile compounds.

0097–6156/92/0490–0140$07.00/0
© 1992 American Chemical Society

EXPERIMENTAL SECTION

Materials.
1) Thiamin hydrochloride (0.5mol) in 1L distilled water (pH 2.8) was heated in an autoclave to 130°C for 2h. Sensory impression: meaty, pungent, sour, rubber, sulfury.
Thiamin hydrochloride (0.5mol) in 1L distilled water (pH 2.8) was heated in an autoclave to 130°C for 6h. Sensory impression: pungent, sour, sulfury.
2) Thiamin hydrochloride (0.9mol) in 1L distilled water (pH adjusted with HCl to 1.5) was heated in an autoclave to 130°C for 6h. Sensory impression: pungent, onion, roasted meat, meaty, processed flavor.
Thiamin hydrochloride (0.9mol) in 1L distilled water (pH adjusted with NaOH to 7.0) was heated in an autoclave to 130°C for 6h. Sensory impression: pungent, butter, diacetyl.
Thiamin hydrochloride (0.9mol) in 1L distilled water (pH adjusted with NaOH to 9.5) was heated in an autoclave to 130°C for 6h. Sensory impression: onion, meaty, caramel, processed flavor.

Isolation of volatiles by simultaneous distillation/extraction.
The respective reaction mixtures were then exposed for 18h to a simultaneous distillation/extraction procedure with a 1:1 mixture of pentane/diethyl ether at atmospheric pressure according to Likens and Nickerson. The resulting extracts were dried over Na_2SO_4 and the organic solvent was carefully removed on a 25cm x 1cm Vigreux distillation column. The concentrated extracts were stored under N_2.

Preseparation by adsorption chromatography.
The respective concentrated flavor extract (pH 9.5) was separated into 6 fractions by medium pressure liquid chromatography on silica gel using a pentane/diethyl ether gradient. The aroma concentrate was placed on a cooled column (480mm x 37mm (i.d.)) filled with 240g of silica gel (25-40µm). The pressure was approximately 2 bar. The flow rate was 10mL/min. All eluates were dried over Na_2SO_4. Each fraction was concentrated to a final volume of 1mL.

Capillary gas chromatography.
Analytical separations were performed on the following three gas chromatographs: Varian 3700, Carlo Erba 5360 Mega Series, and Carlo Erba 4200. The Varian 3700 gas chromatograph was equipped with a modified hot split/splitless injector (Gerstel/Mülheim, Germany) and a commercially available inlet splitter which allowed simultaneous injection into two capillaries of different polarity. The Carlo Erba 5360 gas chromatograph was equipped with the so-called "glass-cap-cross" (Seekamp/Achim, Germany) inlet splitting system. This system for dual column analysis was combined with a cold on-column technique. The Carlo Erba 4200 gas chromatograph was equipped with a flame ionization detector (FID), a flame photometric detector (FPD) operated in the sulfur mode (394nm) and a nitrogen-phosphorus detector (NPD). The column effluent was simultaneously split to the three different detectors using a modified (five "arms" instead of four) glass-cap-cross.
The columns employed were a 60m x 0.32mm (i.d.) DB-WAX fused silica capillary ($d_f = 0.25$µm) and a 60m x 0.32mm (i.d.) DB-1 fused silica capillary ($d_f = 0.25$µm). The columns were obtained from J&W Scientific, Folsom, CA (USA).
The columns were programmed from 60°C to 220°C (3°C/min). Helium carrier gas was used at a flow rate of 2-3 mL/min.
The preparative separations were performed on wide bore DB-1 and DB-WAX fused silica columns (30m x 0.53mm i.d./film thickness 1.0µm - 3.0µm) in an all-glass system. The isolated compounds (low µg-range) were analyzed by capillary GC prior to spectroscopic investigations in order to check their purity.

The quantification was performed by using methyl decanoate as external standard. Obviously, this is not a highly accurate method since the response factors of all compounds were suggested to be 1.0 and no statistical data were calculated. Moreover, the different volatility of each chemical compound in relation to the distillation/extraction was completely neglected. But for the purpose of a rough quantitative picture and most importantly for comparing the different extracts the method proved to be sufficient.

Gas chromatography-mass spectrometry (GC/MS).
A Finnigan MAT Series 8230 (sector field) interfaced by an open split coupling via a flexible heated (250°C) transfer line to a Carlo Erba 5360 Mega Series GC was employed. The operating conditions were as follows: Ion source 220°C, EI 70 eV, cathodic current 1 mA, accelerating voltage 3 kV, resolution 900, scan speed about 1 sec/dec.
A Finnigan MAT Series 112S (sector field) directly coupled to a Varian 3700 Series GC was also used. The operating conditions were as follows: Ion source 230°C, EI 70 eV, cathodic current 0.7 mA, accelerating voltage 800 V, resolution 800, scan speed about 1.3 sec/dec.

Infrared (IR)- and nuclear magnetic resonance (NMR) analysis.
Infrared spectra of isolated samples were obtained in CCl_4 using a Perkin Elmer 983G type instrument.
NMR spectra of collected and of synthesized samples were measured at 400 MHz in $CDCl_3$, C_6D_6 or C_6D_{12} on a Varian VXR-400 instrument with $Si(CH_3)_4$ as internal standard.

Component identification.
Sample components were identified by comparison of the compound's mass spectrum and Kovats index with those of a reference standard. Reference compounds were obtained commercially or synthesized in our laboratory. The respective structures were confirmed by NMR, MS and IR spectroscopies.
Identifications based solely on mass spectral data were considered tentative.

Sensory evaluation.
The olfactory evaluation of selected compounds was performed by an expert panel of flavorists. The synthesized compounds were evaluated in water at certain concentrations. The isolated samples were dissolved in ethanol and tested on a smelling blotter.
In addition, the various extracts were judged for their smell. They were also investigated organoleptically by carrying out GC-sniffing.

RESULTS AND DISCUSSION

The identified flavor compounds of the thermal degradation of thiamin hydrochloride are listed in Tables I-II. Additionally, Figures 1a and 1b show the chemical structures of some of these compounds. Figures 2-4 give proposed routes for the mechanisms of their formation. The peak numbers used for identified compounds correspond to all Tables and Figures in this study. The brackets characterize hypothetical structures, which in most cases are very reactive intermediates or may not be amenable to distillation and gas chromatography.
First, we investigated the influence of the reaction time applying the same tempera- ture (130°C) for 2h and 6h. The results are shown in Table I. The total amount of volatiles determined by capillary GC was 181 mg/mol thiamin (2h) and 240 mg/mol

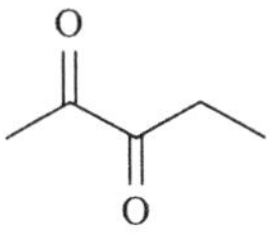

Figure 1a. Various volatile constituents of thermally degraded thiamin.

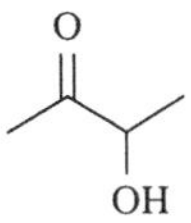

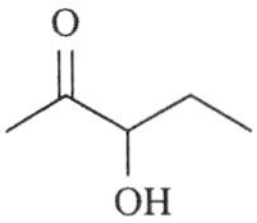

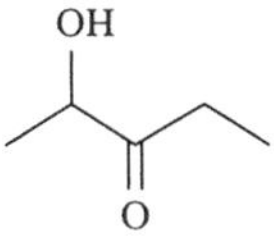

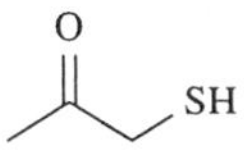

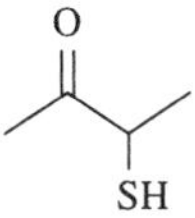

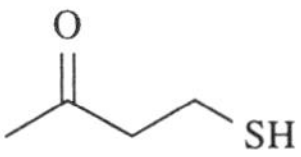

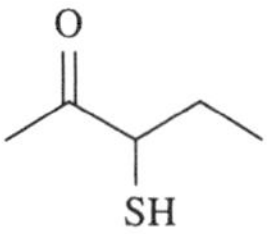

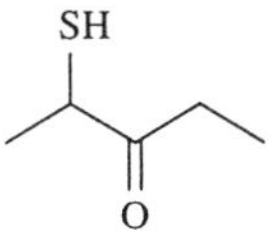

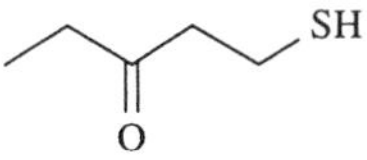

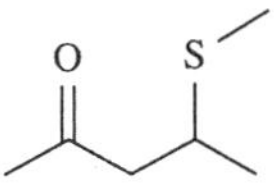

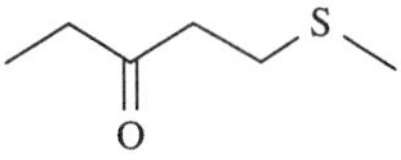

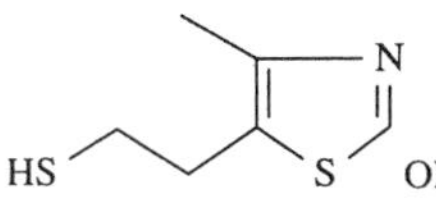

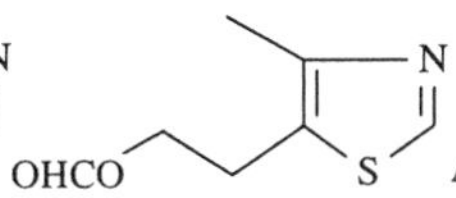

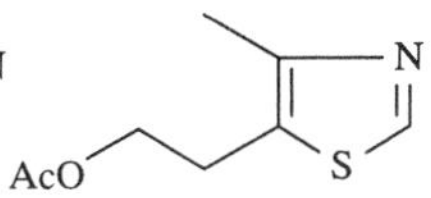

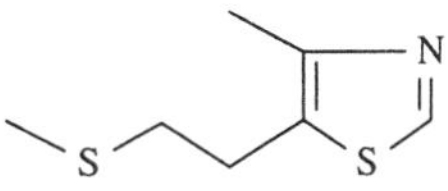

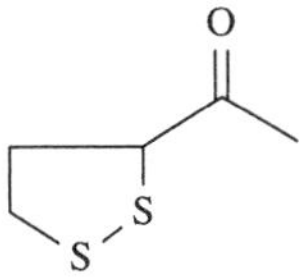

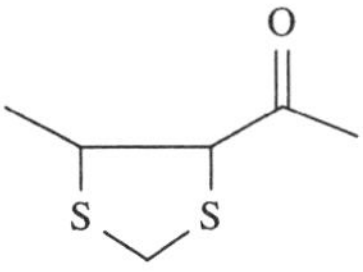

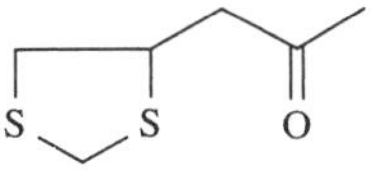

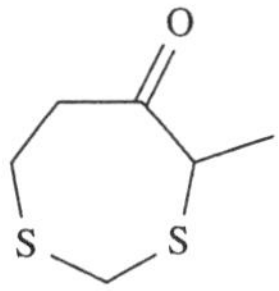

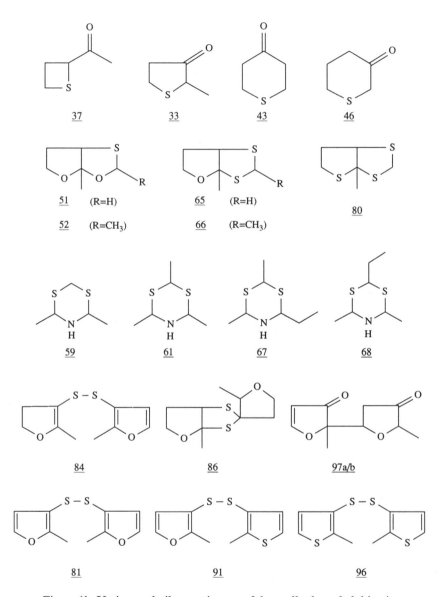

Figure 1b. Various volatile constituents of thermally degraded thiamin.

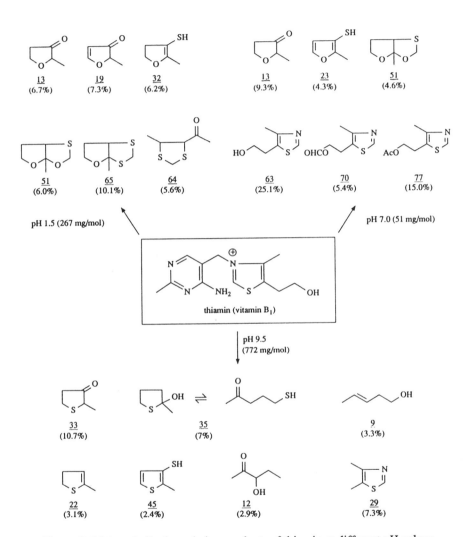

Figure 2. Main volatile degradation products of thiamin at different pH values.

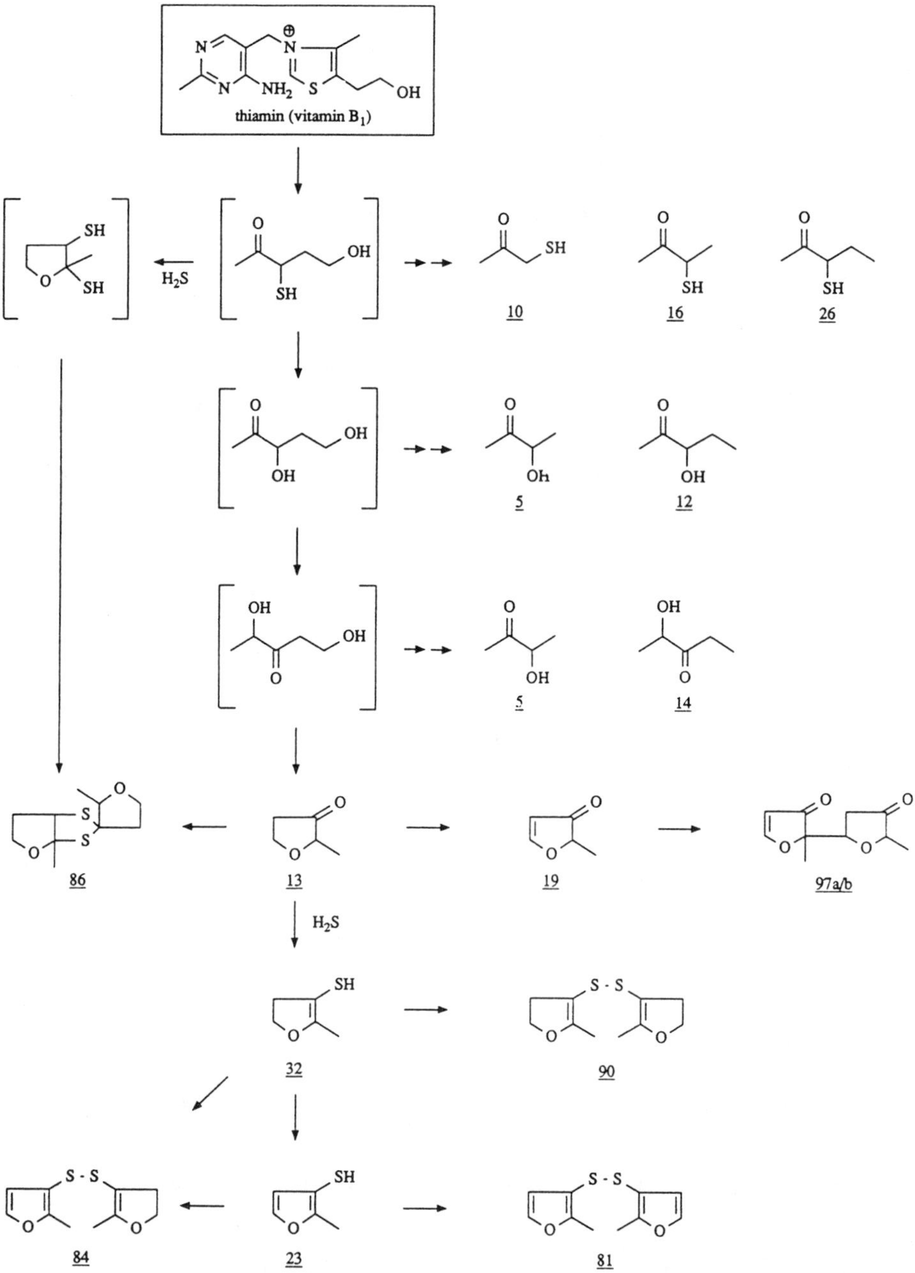

Figure 3. Proposed formation pathways of various thiamin degradation compounds.

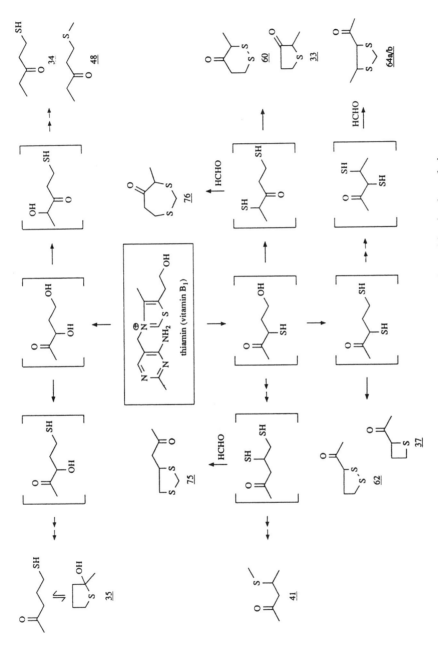

Figure 4. Proposed formation pathways of various thiamin degradation compounds.

Table I. Various constituents of thermally degraded thiamin at different heating times

constituent	peak no.	MW	retention index DB-1	mg/mol thiamin 2 h	mg/mol thiamin 6 h
mercaptopropanone	10	90	739	6.0	1.2
2-methyldihydro-3(2H)-furanone	13	100	782	25.2	47.2
3-mercapto-2-butanone	16	104	787	0.5	0.2
4-methylthiazole	18	99	798	0.2	0.2
2-methyl-3(2H)-furanone	19	98	798	18.0	26.8
2-methyl-3-furanthiol	23	114	844	3.4	2.0
3-mercapto-2-pentanone	26	118	875	0.8	1.8
2-mercapto-3-pentanone	27	118	883	0.1	0.7
2-methyl-4,5-dihydro-3-furanthiol	32	116	919	22.2	8.4
2-methyldihydro-3(2H)-thiophenone	33	116	952	0.3	1.2
2-acetylthietane	37	116	972	0.1	tr.
3-mercaptopropylacetate	42	134	992	2.2	1.7
1-methyl-2,8-dioxa-4-thiabicyclo[3.3.0]octane	51	146	1075	10.6	7.2
1,3-dimethyl-2,8-dioxa-4-thiabicyclo[3.3.0]octane	52	160	1083	1.4	1.1
5-(2-hydroxyethyl)-4-methylthiazole	63	143	1241	0.3	0.4
1,3-dimethyl-2,4-dithia-8-oxabicyclo[3.3.0]octane	66	176	1253	0.3	0.5
chilenone A	97a	196	1364	0.1	2.9
chilenone B	97b	196	1369	0.2	2.2
bis-(2-methyl-3-furyl)disulfide	81	226	1496	tr.	0.2
2-methyl-3-(2-methyl-3-furyldithio)-4,5-dihydrofuran	84	228	1587	0.5	7.4
bis-(2-methyl-4,5-dihydro-3-furyl)disulfide	90	230	1676	15.4	26.4

tr. = trace compound

thiamin (6h). We chose only a few of the major peaks to discuss the differences and to find out how the chemically different compounds behave. Most of them show the expected increase in concentration but some react differently. Primarily the S-compounds with free thiol groups show a decrease. Apparently, they react further to oxidized constituents. A typical example is the ratio of 2-methyl-4,5-dihydro-3-furanthiol 32 and its oxidized dimer bis-(2-methyl-4,5-dihydro-3-furyl)disulfide 90. Of course, this important thiol also reacts to other S-containing volatiles, e.g. 2-methyl-3-(2-methyl-3-furyldithio)-4,5-dihydrofuran 84. Two very important flavor precursors are 2-methyldihydro-3(2H)-furanone 13 and 2-methyl-3(2H)-furanone 19. They are among the compounds with the highest concentrations. Apparently, the latter reacts to a cis/trans mixture of its dimer 97a/b (see Figure 3) which has been named Chilenone A by workers who identified this constituent in the marine alga *Laurencia chilensis* [7]. In our study the two diastereoisomers of Chilenone A show the most dramatic increase in concentration between 2h and 6h.

Secondly, we investigated the influence of different pH values. Table II depicts all flavor compounds identified in this study at pH values of 1.5, 7.0 and 9.5. They are listed along with their peak numbers, molecular weights, Kovats retention indices on a polymethylsiloxane fused silica capillary, their status of identification, and the quantities (mg/mol thiamin) in which they were found in the respective mixture. Some trace compounds are also shown in Table II. They were identified among the volatiles of the respective six silica gel fractions of the alkaline mixture (pH 9.5). Correspondingly, Figure 5 shows S-chromatograms (FPD) of the acidic (top), neutral (middle), and the basic (bottom) thermal degradation of thiamin. These chromatograms demonstrate very clearly, that the degradation under basic conditions leads to the most complex mixture with the highest content of flavor compounds. The yields were 267 mg/mol thiamin (pH 1.5), 51 mg/mol thiamin (pH 7.0), and 772 mg/mol thiamin (pH 9.5). Figure 2 presents the most prominent compounds generated in the respective degradation mixtures. Apparently, not only the yield (quantitative) but also the composition (qualitative) is completely different. Three constituents 5-(2-hydroxyethyl)-4-methylthiazole 63, the so-called sulfurol (25.1%), and two compounds which are related to this basic structure, the formate 70 (5.4%) and acetate 77 (15.0%) comprised nearly 50% of the pH 7.0 reaction mixture. Surprisingly, the two esters of sulfurol are mentioned only once as flavor compounds [Tressl, R., Techn. University of Berlin, Germany, personal communication, 1983]. Evidently, the cleavage of the methylene bond between the pyrimidine and the thiazole ring systems in the thiamin molecule plays the most important role for the thermal degradation at neutral pH. The resulting compounds are 4-amino-5-(hydroxymethyl)-2-methylpyrimidine and 5-(2-hydroxyethyl)-4-methylthiazole 63. Other prominent compounds are 2-methyldihydro-3(2H)-furanone 13 (9.3%), 1-methyl-2,8-dioxa-4-thiabicyclo[3.3.0]octane 51 (4.6%), and 2-methyl-3-furanthiol 23 (4.3%).

The second highest yield of flavor compounds was obtained at acidic pH (1.5). The identified compounds demonstrate very clearly, that the cleavage between the two ring systems also plays an important role for the thermal degradation of thiamin in acidic medium. However, there must be other pathways which lead to the formation of interesting flavor compounds. The major substances are 1-methyl-2,4-dithia-8-oxabicyclo[3.3.0]octane 65 (10.1%), 2-methyl-3(2H)-furanone 19 (7.3%), 2-methyldihydro-3(2H)-furanone 13 (6.7%), 2-methyl-4,5-dihydro-3-furanthiol 32 (6.2%), 1-methyl-2,8-dioxa-4-thiabicyclo[3.3.0]octane 51 (6.0%), and cis/trans-4-acetyl-5-methyl-1,3-dithiolane 64a/b (5.6%). The last one is a new compound not yet cited in the literature. Its formation pathway is postulated in Figure 4 while mass spectral data of one isomer (64a) is shown in Table IV.

The highest yield of flavor compounds was obtained at alkaline pH (9.5). Apparently, this type of degradation takes place by many different and complex ways since most of the identified substances could not be detected in the other two mixtures. The

Table II. Volatile constituents of thermally degraded thiamin at different pH values

constituent	peak no.	MW	identi- fication status	retention index DB-1	mg/mol thiamin		
					pH 1.5	pH 7.0	pH 9.5
2-methyl-4,5-dihydrofuran	1	84	a	670	4.6	–	7.8
2,3-pentanedione	2	100	a	673	–	–	n.q.
2-pentanone	3	86	a	676	3.7	–	n.q.
3-pentanone	4	86	a	690	2.3	–	17.0
acetoine	5	88	a	693	–	–	22.4
3-penten-2-one	6	84	a	710	1.8	0.1	n.q.
1,1-ethanedithiol	7	94	a	712	–	–	tr.
dimethyldisulfide	8	94	a	733	tr.	tr.	tr.
3Z-penten-1-ol	9a	86	a	738	–	tr.	9.4
mercaptopropanone	10	90	a	739	–	tr.	–
3E-penten-1-ol	9b	86	a	750	–	tr.	15.7
2-methylthiophene	11	98	a	766	0.4	tr.	1.1
3-hydroxy-2-pentanone	12	102	c	782	tr.	n.q.	19.6
2-methyldihydro-3(2H)-furanone	13	100	a	782	17.9	4.8	–
2-hydroxy-3-pentanone	14	102	c	784	tr.	0.1	7.0
1-methylthiomethanethiol	15	94	a	784	tr.	–	–
3-mercapto-2-butanone	16	104	a	787	tr.	tr.	–
2,4,5-trimethyl-3-oxazoline	17	113	a	790	–	–	2.8
4-methylthiazole	18	99	a	798	0.2	0.9	–
2-methyl-3(2H)-furanone	19	98	a	798	19.3	1.0	–
2,4,5-trimethyloxazole	20	111	a,g	822	–	–	tr.
1-methylthioethanethiol	21	108	a	826	–	–	tr.
2-methyl-4,5-dihydrothiophene	22	100	a	830	0.2	0.1	24.1
2-methyl-3-furanthiol	23	114	a	844	10.0	2.2	–
4-mercapto-2-butanone	24	104	a	861	–	–	0.8
2-methylenedihydro-3H-thiophene	25	100	a	867	–	–	7.7
3-mercapto-2-pentanone	26	118	a	875	2.2	0.3	–
2-mercapto-3-pentanone	27	118	a	883	0.6	tr.	–
4,5-dimethylthiazole	29	113	a	909	–	0.2	56.1
4,5-dimethyl-2-ethyloxazole	30	125	a	909	–	tr.	–
dihydro-3(2H)-thiophenone	31	102	a	916	–	–	tr.
2-methyl-4,5-dihydro-3-furanthiol	32	116	d	919	16.4	0.9	–
2-methyldihydro-3(2H)-thiophenone	33	116	a	952	1.0	0.2	82.7
1-mercapto-3-pentanone	34	118	a	952	1.4	–	–
2-methyl-2-hydroxytetrahydrothiophene	35	118	a,e	964	–	–	50
⇌ 5-mercapto-2-pentanone	35	118	a,f	964	–	–	50
2,4,5-trimethyl-3-thiazoline	36a	129	a	969	–	–	n.q.
2-acetylthietane	37	116	b	972	–	tr.	–
2,4,5-trimethyl-3-thiazoline	36b	129	a	976	–	–	2.0
2,4,5-trimethylthiazole	38	127	a	980	–	–	0.1
2,5-dimethyldihydro-3(2H)-thiophenone	39	130	a,g	987	–	–	tr.
5-ethyl-4-methylthiazole	40	127	a	990	–	tr.	0.6
4-methylthio-2-pentanone	41	132	a,e	992	–	–	4.4
3-mercaptopropylacetate	42	134	b	992	0.9	–	–
4-thianone	43	116	a	1017	–	–	0.4
2-methyl-2-tetrahydrothiophenethiol	44	134	a	1025	–	–	n.q.
2-methyl-3-thiophenethiol	45	130	a	1030	–	–	18.4
3-thianone	46	130	a	1037	–	–	13.2
4-oxopentylacetate	47	144	a,f	1046	1.1	tr.	–
1-methylthio-3-pentanone	48	132	a	1053	1.6	n.q.	16.3
tetramethylpyrazine	49	136	a	1069	–	–	22.0
2-methyl-3-tetrahydrothiophenethiol	50a	134	a,g	1069			tr.
1-methyl-2,8-dioxa-4-thiabicyclo[3.3.0]octane	51	146	a	1075	16.0	2.3	–
1,3-dimethyl-2,8-dioxa-4-thiabicyclo[3.3.0]octane	52	160	a	1083	0.6	tr.	–
2-methyl-3-tetrahydrothiophenethiol	50b	134	a,g	1100			tr.

Table II. continued

constituent	peak no.	MW	identi-fication status	retention index DB-1	mg/mol thiamin		
					pH 1.5	pH 7.0	pH 9.5
3,5-dimethyl-1,2,4-trithiolane	53	152	a	1102	—	—	0.1
tetramethyl-1H-pyrrole	54	123	a	1109	—	—	1.2
2-methylthio-5-methylthiophene	56	144	a,g	1117			tr.
2-ethyl-3,5,6-trimethylpyrazine	57	152	a	1138	—	—	3.4
kahweofuran	58	140	a	1160	—	—	0.9
4,6-dimethyl-5,6-dihydro-1,3,5-dithiazine	59	149	a,g	1162			tr.
3-methyl-1,2-dithian-4-one	60	148	b	1168	tr.	—	—
thialdine	61	163	a	1172	—	—	10.1
3-acetyl-1,2-dithiolane	62	148	a	1199	—	0.2	
5-(2-hydroxyethyl)-4-methylthiazole	63	143	a	1241	0.4	12.9	1.1
4-acetyl-5-methyl-1,3-dithiolane	64a	162	b,e	1245	14.9	—	—
1-methyl-2,4-dithia-8-oxabicyclo[3.3.0]octane	65	162	a	1245	27.0	tr.	1.6
1,3-dimethyl-2,4-dithia-8-oxabicyclo[3.3.0]octane	66	176	b	1253	1.7	—	—
2,6-dimethyl-4-ethyl-5,6-dihydro-1,3,5-dithiazine	67	177	a	1261	—	—	n.q.
4-acetyl-5-methyl-1,3-dithiolane	64b	162	b,e	1265	1.0	—	1
4,6-dimethyl-2-ethyl-5,6-dihydro-1,3,5-dithiazine	68	177	a	1272	—	—	0.3
5-(2-formyloxyethyl)-4-methylthiazole	70	171	a	1275	0.7	2.8	—
2-methyl-2-acetylthiotetrahydrothiophene	71	176	a,e,g	1275			tr.
5-(2-methyl-2-tetrahydrofuryloxy)-2-pentanone	72	186	a,f	1284	0.4	—	n.q.
5-(2-mercaptoethyl)-4-methylthiazole	73	159	a	1284	—	—	n.q.
2-methyl-3-methyldithiothiophene	74	176	a,g	1330			tr.
4-(2-oxopropyl)-1,3-dithiolane	75	162	b,e	1332	—	—	8.3
4-methyl-1,3-dithiepan-5-one	76	162	b,e	1340	6.1	—	—
5-(2-acetoxyethyl)-4-methylthiazole	77	185	a	1352	0.1	7.7	—
5-(2-methylthioethyl)-4-methylthiazole	79	173	a	1383	—	—	1.9
1-methyl-2,4,8-trithiabicyclo[3.3.0]octane	80	178	c	1415	tr.	—	0.2
bis-(2-methyl-3-furyl)disulfide	81	226	a	1496	—	tr.	—
2-methyl-3-(2-methyl-2-tetrahydrothienylthio)furan	83	214	a,g	1541			tr.
2-methyl-3-(2-methyl-3-furyldithio)-4,5-dihydrofuran	84	228	b	1587	1.4	0.1	—
2-methyl-3-(2-methyl-3-tetrahydrothienylthio)furan	85	214	a,g	1610			tr.
2',3a-dimethylhexahydrospiro-[1,3-dithiolo[4,5-b]furan-2,3'-(2'H)furan]	86	232	b,g	1634			tr.
2-methyl-3-(2-tetrahydrothienylmethylthio)furan	87	214	a,g	1651			tr.
5-(2-methyl-2-tetrahydrothienylthio)-2-pentanone	89	218	a,f	1672	—	—	2.4
bis-(2-methyl-4,5-dihydro-3-furyl)disulfide	90	230	a	1676	7.8	0.7	—
2-methyl-3-(2-methyl-3-thienyldithio)furan	91	242	a,g	1681			tr.
2-methyl-2-(2-methyl-3-thienylthio)tetrahydrothiophene	92	230	a	1731	—	—	1.5
2-methyl-3-(2-methyl-3-thienylthio)tetrahydrothiophene	94	230	a	1805	—	—	0.4
2-methyl-3-(2-tetrahydrothienylmethylthio)thiophene	95	230	a,g	1848			tr.
bis-(2-methyl-3-thienyl)disulfide	96	258	a	1868	—	—	2.9

[a]Identified by comparison of the compound's mass spectrum and retention index with those of an authentic reference standard.

[b]Not synthesized but identified by means of the spectra (NMR, IR,MS) of the isolated sample.

[c]Tentatively identified by comparison with a mass spectrum reported in the literature.

[d]Tentatively identified by interpretation of the mass spectrum.

[e]Not previously mentioned in the literature.

[f]Cited in the literature but not in relation to flavor chemistry.

[g]Identified only in one of the six silica gel fractions (pH 9,5).

tr.= trace compound

n.q.= not quantified because of overlapping peaks.

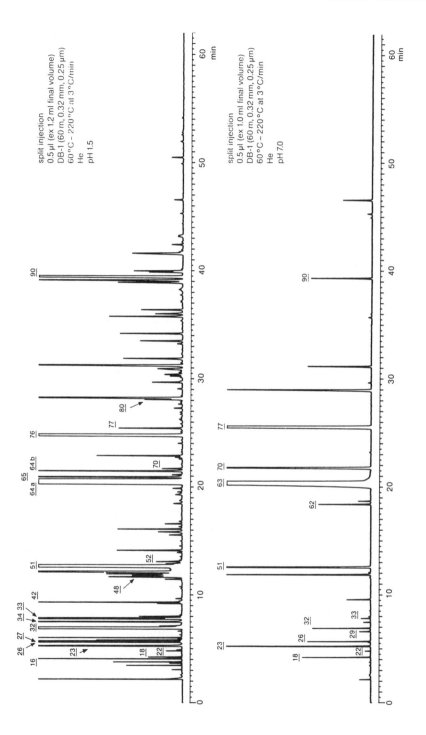

split injection
0.5 µl (ex 1.2 ml final volume)
DB-1 (60 m, 0.32 mm, 0.25 µm)
60 °C – 220 °C at 3 °C/min
He
pH 1.5

split injection
0.5 µl (ex 1.0 ml final volume)
DB-1 (60 m, 0.32 mm, 0.25 µm)
60 °C – 220 °C at 3 °C/min
He
pH 7.0

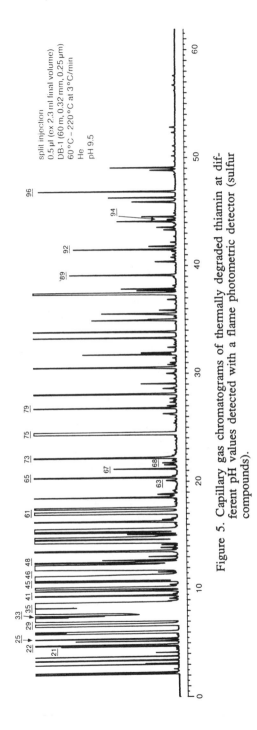

Figure 5. Capillary gas chromatograms of thermally degraded thiamin at different pH values detected with a flame photometric detector (sulfur compounds).

variety of structures is much greater. The main compounds are 2-methyldihydro-3(2H)-thiophenone 33 (10.7%), 4,5-dimethylthiazole 29 (7.3%), 2-methyl-2-hydroxytetrahydrothiophene 35 (7%), E/Z-3-penten-1-ol 9a/b (3.3%), 2-methyl-4,5-dihydrothiophene 22 (3.1%), 3-hydroxy-2-pentanone 12 (2.9%), and 2-methyl-3-thiophenethiol 45 (2.4%). It is striking in particular, that the degradation of thiamin at alkaline pH is very severe and leads to the formation of hydrogen sulfide, ammonia, formic acid, formaldehyde, acetaldehyde, and acetic acid in relatively high amounts. This is evident through the presence of many known flavor compounds whose formation seems only possible from these basic chemicals, e.g. 2,4,5-trimethyl-3-oxazoline 17, 2,4,5-trimethyloxazole 21, 4,5-dimethyl-2-ethyloxazole 30, tetramethylpyrazine 49, 2-ethyl-3,5,6-trimethylpyrazine 57, tetramethyl-1H-pyrrole 54, 2,4,6-trimethyl-5,6-dihydro-1,3,5-dithiazine (thialdine) 61, 4,6-dimethyl-5,6-dihydro-1,3,5-dithiazine 59, 2-ethyl-4,6-dimethyl-5,6-dihydro-1,3,5-dithiazine 68, 4-ethyl-2,6-dimethyl-5,6-dihydro-1,3,5-dithiazine 67, 3,5-dimethyl-1,2,4-trithiolane 53, cis/trans-2,4,5-trimethyl-3-thiazoline 36a/b, and 2,4,5-trimethylthiazole 38.
Additionally, sulfur plays a more important role at alkaline pH than in the mixtures with acidic and neutral pH. In general, many more thiophenes are formed at alkaline pH while the furans are missing. The most striking example is shown by 2-methyl-3-thiophenethiol 45 (18.4 mg/mol at pH 9.5 and n.d. at pH 1.5) compared to 2-methyl-3-furanthiol 23 (n.d. at pH 9.5 and 10.0 mg/mol at pH 1.5). We identified many S-containing heterocyclic compounds which were not detected or were present in lesser amounts in the other two extracts. Examples are 2-methyl-4,5-dihydrothiophene 22, 2-methylene-4,5-dihydro-3H-thiophene 25, 2-methyldihydro-3(2H)-thiophenone 33, 3-thianone 46, 5-(2-mercaptoethyl)-4-methylthiazole 73, 5-(2-methylthioethyl)-4-methylthiazole 79, bis-(2-methyl-3-thienyl)disulfide 96, and 4-(2-oxopropyl)-1,3-dithiolane 75. The last one is a new compound not known in literature thus far. Its probable formation is depicted in Figure 4. Its mass and NMR spectra are shown in Table IV and Figure 7, respectively, while the sensory impression is described in Table III. Remarkably, 5-(2-mercaptoethyl)-4-methylthiazole 73 and 5-(2-methylthioethyl)-4-methylthiazole 79 are mentioned only once as flavor compounds [Tressl, R., Techn. University of Berlin, Germany, personal communication, 1983]. Mass spectral data of both compounds are given in Table IV while the sensory impressions are described in Table III. Noteworthy is the well-known coffee flavor compound kahweofuran 58 [8]. Its formation from pure thiamin is not clear but it underlines the generation of small chemical units at the alkaline degradation once again. Possibly, kahweofuran 58 is generated from mercaptopropanone by a pathway similar to that described by Tressl [9].
The whole thiamin project shows strikingly that many compounds were identified which are known as products from the thermal reaction of sugars and amino acids (e.g. xylose + cysteine) [see for example 10]. Apparently, one of the main degradation products of thiamin, 5-hydroxy-3-mercapto-2-pentanone, reacts in a way like a C-5 sugar either with itself or with other generated intermediates. Therefore, in a reaction mixture containing xylose, cysteine (cystine), and thiamin some of the compounds are probably formed on two different ways (e.g. 2-methyldihydro-3(2H)-furanone 13). It is also obvious now that the heterocyclic thioethers identified by us for the first time in a reaction mixture containing thiamin, cystine, monosodium glutamate, and ascorbic acid [2,3] are, at least partially, thiamin degradation products. We were able to trace all these interesting constituents in our present study. Examples are 2-methyl-3-(2-methyl-2-tetrahydrothienylthio)furan 83, 2-methyl-3-(2-methyl-3-tetrahydrothienylthio)furan 85, 2-methyl-3-(2-tetrahydrothienylmethylthio)furan 87, 2-methyl-2-(2-methyl-3-thienylthio)tetrahydrothiophene 92, 2-methyl-3-(2-methyl-3-thienylthio)tetrahydrothiophene 94, and 2-methyl-3-(2-tetrahydrothienylmethylthio)-thiophene 95. Interestingly, most of these heterocyclic thioethers could only be identified in the alkaline (pH 9.5) thermal degradation of thiamin while in our former reaction mixture [2,3] the pH was 5.0. The reason for this result might be the release

Table III. Sensory impressions of some selected thiamin degradation compounds

constituent	peak no.	isolated sample[a] mg/100 mg	synthesized sample[b] ppm	sensory impression
4-mercapto-2-butanone	24		3.75×10^{-2}	meaty, sulfury, roasty, potato, lard
1-mercapto-3-pentanone	34		1.13×10^{-1}	rubber, durian, onion, sulfury
2-methyl-2-hydroxytetrahydrothiophene	35		7.5×10^{-2}	tropical fruit, fruity, sulfury, meaty, black currant
4-methylthio-2-pentanone	41		7.5×10^{-2}	potato, cauliflower, broccoli, vegetables, herbal
1-methylthio-3-pentanone	48		2.25	toffee
2-methyl-2-acetylthiotetrahydrothiophene	71		7.5×10^{-1}	onion, malt, roasty, meaty, peanut
5-(2-mercaptoethyl)-4-methylthiazole	73		3×10^{-2}	tuna, metallic, sulfury, tropical fruit, thiamin
4-(2-oxopropyl)-1,3-dithiolane	75	5		cabbage, sulfury, vegetables
4-methyl-1,3-dithiepan-5-one	76	5		meaty, mushroom, sweet, metallic
5-(2-methylthioethyl)-4-methylthiazole	79		3×10^{-2}	fatty, metallic, sulfury, chicken

[a]dissolved in ethanol and tested on a smelling blotter
[b]tested in water

Table IV. Mass spectral data of selected thiamin degradation compounds

constituents	peak no.	m/e (%)				
2-methyl-2-hydroxytetrahydrothiophene	35	43 (100)	58 (56)	71 (37)	59 (31)	118 (24)
⇌ 5-mercapto-2-pentanone		41 (20)	47 (16)	39 (14)	60 (13)	85 (12)
4-methylthio-2-pentanone	41	43 (100)	132 (52)	89 (33)	75 (27)	41 (22)
		85 (14)	117 (9)	47 (7)	49 (6)	61 (5)
4-acetyl-5-methyl-1,3-dithiolane	64a	43 (100)	119 (88)	162 (87)	85 (81)	115 (72)
		73 (66)	45 (38)	74 (28)	59 (26)	75 (22)
2-methyl-2-acetylthiotetrahydrothiophene	71	101 (100)	59 (31)	43 (27)	100 (26)	67 (19)
		99 (18)	85 (18)	45 (10)	41 (9)	39 (8)
5-(2-mercaptoethyl)-4-methylthiazole	73	112 (100)	45 (29)	85 (29)	113 (26)	159 (19)
		132 (13)	47 (8)	39 (7)	59 (7)	27 (6)
4-(2-oxopropyl)-1,3-dithiolane	75	104 (100)	43 (82)	162 (33)	73 (23)	45 (17)
		103 (12)	105 (11)	106 (9)	74 (6)	119 (6)
4-methyl-1,3-dithiepan-5-one	76	74 (100)	162 (100)	60 (76)	46 (34)	45 (23)
		59 (19)	41 (17)	27 (16)	106 (15)	55 (15)
5-(2-methylthioethyl)-4-methylthiazole	79	61 (100)	112 (35)	173 (22)	45 (14)	125 (12)
		85 (10)	27 (6)	35 (5)	63 (5)	59 (4)
2',3a-dimethylhexahydrospiro-[1,3-dithiolo[4,5-b]furan-2,3'(2'H)furan]	86	84 (100)	43 (46)	83 (22)	150 (9)	116 (8)
		85 (7)	73 (6)	232 (5)	71 (5)	45 (4)

and the presence of hydrogen sulfide which in the case of pure thiamin is highest at an alkaline pH while in our former reaction mixture cystine served as an additional source of hydrogen sulfide.

We were able to identify a few flavor compounds in this study which to the best of our knowledge are not mentioned in literature so far [11]. Remarkably, 2-methyl-2-hydroxytetrahydrothiophene 35 is one of the major peaks in the alkaline degradation. It seems to be in an chemical equilibrium with 5-mercapto-2-pentanone and is hardly amenable to gas chromatography. Therefore, its content in our mixture with 50 mg/mol is very approximate. While 2-methyl-2-hydroxytetrahydrothiophene 35 is not cited in the literature 5-mercapto-2-pentanone 35 is mentioned once in a japanese pharmaceutical patent but not in relation to flavor chemistry. The formation of these compounds is well explainable from thiamin (see Figure 4). The sensory impression and the mass spectral data are presented in Tables III and IV, respectively. 1-Methyl-2,4,8-trithiabicyclo[3.3.0]octane 80 is not yet cited in the literature. Nevertheless, Tressl and co-workers [Tressl, R., Techn. University of Berlin, Germany, personal communication, 1983] have previously identified this compound as a thiamin degradation product. Its formation mechanism was postulated by us previously [1]. 2-Methyl-2-acetylthiotetrahydrothiophene 71, the thioacetate of 2-methyl-2-tetrahydrothiophenethiol 44, could be identified as a trace component in the alkaline degradation mixture. Its formation shows once again that under the influence of hydrogen sulfide and acetic acid even a thioester can be generated. The sensory impression of 71 is given in Table III while the mass spectral data is shown in Table IV. Moreover, its NMR spectrum is depicted in Figure 6. Two other interesting new compounds are 4-(2-oxopropyl)-1,3-dithiolane 75 and 4-methyl-1,3-dithiepan-5-one 76. Their formation pathways are postulated in Figure 4. Apparently, both are generated from C-5 intermediates by reaction with formaldehyde. Their sensory properties are described in Table III and their mass spectra are shown in Table IV. In addition, their NMR spectra are depicted in Figure 7.

A further category of identified compounds in Tables I and II includes those which are known in literature but not in relation to flavor chemistry. Interesting examples are 5-acetoxy-2-pentanone 47, 5-(2-methyl-2-tetrahydrofuryloxy)-2-pentanone 72, and 5-(2-methyl-2-tetrahydrothienylthio)-2-pentanone 89. These compounds show once again the importance of C-5 intermediates, in this case of 5-hydroxy-2-pentanone and its cyclized form 2-hydroxy-2-methyltetrahydrofuran as well as 5-mercapto-2-pentanone 35 and 2-methyl-2-hydroxytetrahydrothiophene 35.

A chemically remarkable constituent is 2',3a-dimethylhexahydrospiro-[1,3-dithiolo-[4,5-b]furan-2,3'(2'H)furan] 86. It was identified as trace compound among the alkaline degradation products. This tricyclic thioacetal is mentioned only in the patent literature [12]. Our identification is based on spectral data from an isolated sample. But its formation from 2-methyldihydro-3(2H)-furanone 13 and the postulated intermediate 2-methyl-2-mercapto-3-tetrahydrofuranthiol seems plausible (Figure 3). The mass spectrum of 86 is shown in Table IV.

An interesting class of flavor volatiles is represented by mercaptoketones and hydroxyketones. In the present study we were able to identify many different examples of them. Additionally, a few methylthioketones could be found. Some of them are known flavor compounds whose formation is readily explainable by certain degradation steps of thiamin (Figures 2-4), e.g. 3-hydroxy-2-pentanone 12 [13-16], 2-hydroxy-3-pentanone 14 [14-18], 3-mercapto-2-pentanone 26 [1,19-25], 2-mercapto-3-pentanone 27 [19-21], and 3-mercapto-2-butanone 16 [1,19-21,26]. 4-Mercapto-2-butanone 24 is mentioned only once in literature in relation to flavors [27]. In contrast, 1-mercapto-3-pentanone 34 is described as Maillard reaction product [20,22]. The sensory impressions of 24 and 34 are given in Table III. 4-Methylthio-2-pentanone 41 is a new flavor compound. Its sensory properties and spectral data are depicted in Tables III and IV as well as in Figure 6. Interestingly, 1-methylthio-3-pentanone 48 was found so far solely in various vegetables, e.g. green kohlrabi [28],

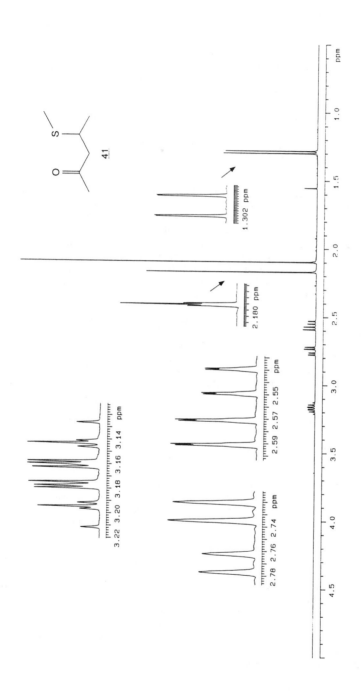

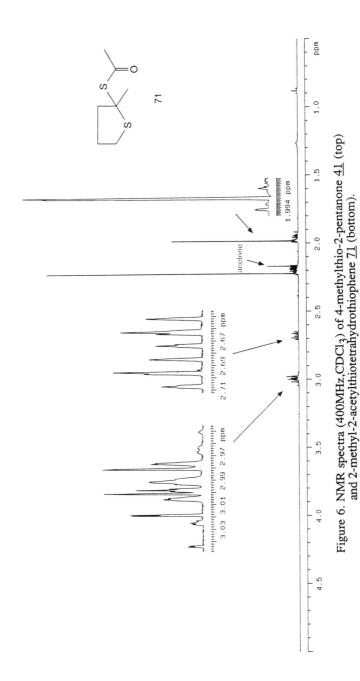

Figure 6. NMR spectra (400MHz,CDCl₃) of 4-methylthio-2-pentanone $\underline{41}$ (top) and 2-methyl-2-acetylthiotetrahydrothiophene $\underline{71}$ (bottom).

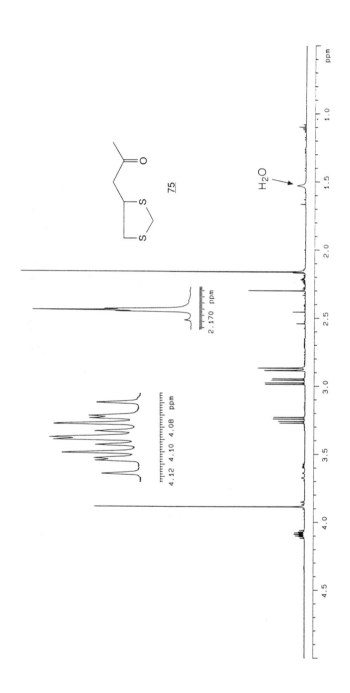

75

H_2O

2.170 ppm

4.12 4.10 4.08 ppm

ppm

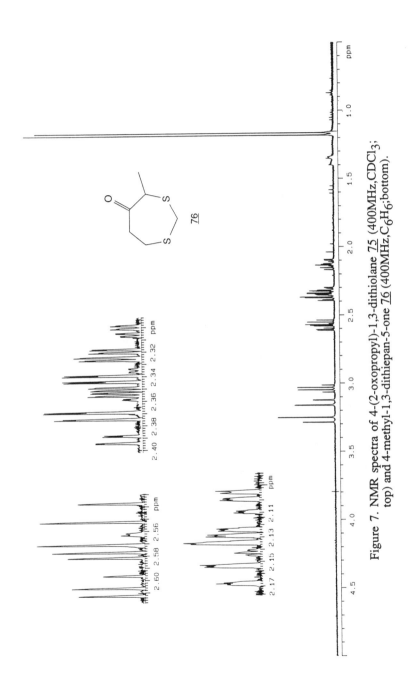

Figure 7. NMR spectra of 4-(2-oxopropyl)-1,3-dithiolane 75 (400MHz,CDCl₃; top) and 4-methyl-1,3-dithiepan-5-one 76 (400MHz,C₆H₆;bottom).

tomato [29], and radish [30]. Its sensory impression is best described as "toffee" (Table III). These last two compounds which were found mainly in the alkaline degradation demonstrate very clearly (Figure 4), that the methylthio-group must be formed from reactive intermediates with free thiol groups, probably in a radical reaction. Moreover, 4-methylthio-2-pentanone 41 may origin from 4,5-dimercapto-2-pentanone which should be the precursor as well for 4-(2-oxopropyl)-1,3-dithiolane 75.

ACKNOWLEDGMENTS

The authors would like to thank their entire groups of organic synthesis, chromatography, spectrometry, and flavor research for their valuable and skillful work. They also express their gratitude to I. Güntert and A. Sagebiel for drawing the chemical structures and supporting the project. Ron Buttery (USDA Albany, California) is thanked for a reference sample of 1-methylthio-3-pentanone.   We dedicate this chapter to Dr. Walter Jennings on the occasion of his 70th birthday.

LITERATURE CITED

1. Güntert, M.; Brüning, J.; Emberger, R.; Köpsel, M.; Kuhn, W.; Thielmann, T.; Werkhoff, P. *J. Agric. Food Chem.* **1990**, *38*, 2027.
2. Werkhoff, P.; Emberger, R.; Güntert, M.; Köpsel, M. "Isolation and Characterization of Volatile Sulfur-Containing Meat Flavour Components in Model Systems". In *Thermal Generation of Aromas*; Parliment, T.H.; Mc Gorrin, R.J.; Ho, C.-T, Eds.; American Chemical Society: Washington D.C., 1989, pp. 460.
3. Werkhoff, P.; Brüning, J.; Emberger, R.; Güntert, M.; Köpsel, M.; Kuhn, W.; Surburg, H. *J. Agric. Food Chem.* **1990**, *38*, 777.
4. Werkhoff, P.; Brüning, J.; Emberger, R.; Güntert, M.; Köpsel, M.; Kuhn, W. In *11th International Congress of Essential Oils, Fragrances and Flavours*, New Delhi, India, 12-16 November 1989; Proceedings Volume 4; Chemistry, Analysis and Structure; Bhattacharyya, S.C., Sen, N., Sethi, K.L., Eds.; Oxford & IBH Publishing CO.PVT.LTD.: New Delhi, India, 1989, pp 215.
5. Werkhoff, P.; Bretschneider, W.; Emberger, R.; Güntert, M.; Hopp, R.; Köpsel,M. *Chem. Mikrobiol. Technol. Lebensm.* **1991**, *13*, 30.
6. Werkhoff, P.; Bretschneider, W.; Güntert, M.; Hopp, R.; Köpsel, M; Surburg, H. *Chem. Mikrobiol. Technol. Lebensm.* **1991**, *13*, 111.
7. San Martin, A.; Rovirosa, J.; Munoz, O.; Chen,M.H.M.; Guneratne,R.D.; Clardy,J. *Tetrahedron Lett.* **1983**, *24*, 4063.
8. Flament, I.; Chevallier, C. *Chemistry and Industry* **1988**, 592.
9. Tressl, R. In *Flavour Science and Technology*; Bessiere, Y., Thomas, A.F., Eds.; John Wiley & Sons: Chichester, England, 1990, pp 87.
10. Shahidi, F.; Rubin, L.J.; D'Souza, L.A. *CRC Crit. Rev. Food Sci. Nutr.* **1986**, *24*, 141.
11. Searched up to *Chem. Abstr.* **1990**, *113*.
12. Neth. Patent 7 100 235, 1972.
13. Fröhlich, O.; Schreier, P. *Flavour Fragrance J.* **1989**, *4*, 177.
14. Roedel, W.; Habisch, D. *Nahrung* **1989**, *33*, 449.
15. Shu, C.K.; Ho, C.-T. *J. Agric. Food Chem.* **1988**, *36*, 801.
16. Kunert-Kirchhoff, J.; Baltes, W. *Z. Lebensm.-Unters. Forsch.* **1990**, *190*, 14.
17. Silvar, R.; Kamperschroer, H.; Tressl, R. *Chem. Mikrobiol. Technol. Lebensm.* **1987**, *10*, 176.
18. Shu, C.K.; Hagedorn, M.L.; Ho, C.-T. *J. Agric. Food Chem.* **1986**, *34*, 344.
19. Misharina, T.A.; Golovnya, R.V. *Abh. Akad. Wiss. DDR, Abt. Math., Naturwiss., Tech.* **1988**, 115.

20. Farmer, L.J.; Mottram, D.S. *J. Sci. Food Agric.* **1990**, *53*, 505.
21. Misharina, T.A.; Vitt, S.V.; Golovnya, R.V. *Biotekhnologiya* **1987**, *3*, 210.
22. Farmer, L.J.; Mottram, D.S.; Whitfield, F.B. *J. Sci. Food Agric.* **1989**, *49*, 347.
23. Gasser, U.; Grosch, W. *Z. Lebensm.-Unters. Forsch.* **1990**, *190*, 511.
24. Gasser, U.; Grosch, W. *Z. Lebensm.-Unters. Forsch.* **1990**, *190*, 3.
25. Shu, C.K.; Hagedorn, M.L.; Mookherjee, B.D.; Ho, C.-T. *J. Agric. Food Chem.* **1985**, *33*, 638.
26. Okumara, J.; Yamai, T.; Yajima, J.; Hayashi, K. *Agric. Biol. Chem.* **1990**, *54*, 1631.
27. U.S. Patent 4 464 408, 1983.
28. MacLeod, G.; MacLeod, A.J. *Phytochemistry* **1990**, *29*, 1183.
29. Buttery, R.G.; Teranishi, R.; Flath, R.A.; Ling, L.C. *J. Agric. Food Chem.* **1990**, *38*, 792.
30. Kjaer, A.; Madsen, J.O.; Maeda, Y.; Ozawa, Y.; Uda,Y. *Agric. Biol. Chem.* **1978**, *42*, 1715.

RECEIVED December 18, 1991

Chapter 12

Formation of Furaneol in Heat-Processed Foods

Peter Schieberle

Deutsche Forschungsanstalt für Lebensmittelchemie, Lichtenbergstrasse 4, D–8046 Garching, Germany

2,5-dimethyl-4-hydroxy-3(2H)-furanone (Fura-neol) occurs in many processed foods and was recently established as important odorant in wheat bread crust and popcorn. To gain an insight into its origin in such foodstuffs, the precursors in wheat dough and corn were elucidated on the basis of quantitative measurements. Baking experiments revealed yeast as the most important source of Fura-neol in wheat bread crust. Furthermore, boiling of an aqueous solution of the low molecular weight compounds from disrupted yeast cells yielded high amounts of Furaneol. Analysis of the free sugars present in this fraction and results of model experiments indicated fructose-1,6-di-phosphate as the predominant precursor of the odorant. In contrast, dry-heating was necessary to liberate Furaneol from its precursors in corn. Model experiments showed that these conditions, simulating the popping process, caused the formation of Furaneol from glucose and fructose which predominated in corn and were little active, when heated in aqueous solution. Further model experiments showed that acetylformoine is an important intermediate in the formation of Furaneol from hexoses.

In the flavor profiles of heat-processed foods containing carbohydrates, caramel-like odor notes frequently contribute to the overall flavor impressions. Among the flavor compounds exhibiting caramel-like odors, 2,5-dimethyl-4-hydroxy-3(2H)-furanone (Furaneol) is of special interest because of its relatively low flavor thresholds

0097–6156/92/0490–0164$06.00/0

© 1992 American Chemical Society

of 0.03 mg (taste) and 0.1 mg per liter water (odor)
[1].
The odorant was reported for the first time in pineapples [2] and strawberries [3]. Since a glucoside of
the odorant has been identified recently [4], it seems
likely that in the fruits a biogenetic pathway is involved in its formation.
Furaneol has been detected in several heat-processed
foods, e.g. beef broth [5], roasted almonds [6] or
roasted coffee [7] and was recently identified as
important odorant in wheat bread crust [8] and popcorn
[9].
Model experiments have revealed that the compound is
formed by a thermal degradation of 6-deoxy sugars, e.g.
rhamnose [10-12], and also of fructose [13] in the presence of amines or amino acids.
To gain an insight into the origin of the flavor compound in heat-processed foods, the precursors of Furaneol in wheat bread and popcorn were elucidated. Two series of experiments were performed: first, studies to
localize the precursors in wheat dough and corn were undertaken and, secondly, the formation of Furaneol was
elucidated in model experiments.

Results

Quantification of Furaneol. Quantitative measurements
are a prerequisite to study the formation of odorants in
complex food systems. Due to its relatively high polarity and its instability under normal gas chromatographic
conditions, the quantification of Furaneol may lead to
incorrect results. To overcome these problems, we have
recently developed a stable isotope dilution assay
(SIDA) on the basis of carbon-13 labelled Furaneol [14].
In a first experiment the SIDA was applied to extracts
of wheat bread crust and popcorn. As shown in Table I,
relatively high concentrations of Furaneol were present
in both foods. Furthermore, a comparison of the odor
units (OU) with those of 2-acetyl-1-pyrroline, previously established as one of the most important odorants in both flavors [8, 9], underlined the sensory importance of Furaneol in the flavors of both foodstuffs.

Precursors in wheat dough. To gain a first insight into
the source of the precursors, the amount of Furaneol
present in bread crusts prepared from either a yeasted
dough and a model dough in which yeast had been replaced
by a leavening agent was determined. The results revealed (Table II) that substitution of the yeast by the
leavening agent led to a significant decrease of the
Furaneol concentration in the crust. The data indicated
that nearly 90 % of the Furaneol present in the bread
crust prepared from the yeasted dough were liberated
from the yeast.

Table I: Concentrations and odor units (OU)[a] of Furaneol
and 2-acetyl-1-pyrroline in wheat bread crust
and popcorn

Odorant	Wheat bread crust		Popcorn	
	µg/kg	OU	µg/kg	OU
Furaneol	1960	1960	1370	1370
2-Acetyl-1-pyrroline	96	4800	32	1600

[a] The OU were calculated by dividing the concentrations
of the odorants by their odor thresholds (Furaneol: 1
ng/L air; 2-acetyl-1-pyrroline: 0.02 ng/L air). Odor
thresholds were approximated as described in [9].

Table II. Influence of bakers yeast on the concentrations of Furaneol in wheat bread crust[a]

Crust obtained from a	Furaneol (µg/kg crust)
yeasted dough	1960
chemically leavened dough	218

[a] The dough was prepared from 500 g wheat flour (type
550), 15 g NaCl, 10 g sucrose and 270 ml of tap water.
Either yeast (30 g) or a commercial leavening agent
was added.

To test whether the flavor compound was already
present in the yeast or whether it was liberated from
precursors during the baking process, the following
experiment was performed: Yeast cells were disrupted by
grinding with silica gel in a mortar. The resulting
material was extracted with diethyl ether and analyzed
for Furaneol. It was found that the yeast already
contained 871 µg/kg of the flavor compound. The ether-
extracted yeast cell/silica material was then heated (15
min; 170°C) and the Furaneol formed was determined.
Compared to the non-heated yeast the amounts increased
by a factor of more than ten, indicating that the major
part of Furaneol present in bread crust was supplied by
a degradation of precursors in yeast. To locate these
precursors, yeast cells were disrupted in a cell mill
and fractionated by centrifugation and ultrafiltration
as described recently [8]. By heat treatment of each of
the fractions obtained, the precursors of Furaneol could
be located in the fraction of the water soluble, low-
molecular weight compounds (LMW-fraction).
In two different model systems the influence of the
heating conditions on the amounts of Furaneol liberated
from the LMW-fraction were then studied. As shown in Ta-

Table III. Formation of Furaneol from the fraction of water-soluble, low molecular weight compounds of yeast

Expt.	Heating conditions		Furaneol
	Time (min)	Temp. (°C)	(μg/kg yeast)
1[a]	60	100	7825
2[b]	45	150	13071

[a] The material (from 5 g of yeast), dissolved in phosphate buffer (20 ml; 0.1 mol/L; pH 7.0) was heated under reflux.

[b] cf. footnote a. Heating was performed in a laboratory autoclave.

Table IV. Formation of Furaneol from carbohydrates - Influence of reaction time and temperature[a]

Carbohydrate (0.2 mmol)	Furaneol (μg)	
	60 min (100°C)	45 min (150°C)
Sucrose	<0.1	<0.1
Glucose	<0.1	0.9
Fructose	0.6	2.9
Rhamnose	20.3	282.0
Rhamnose	n.a.	6.6[b]

[a] The carbohydrate, dissolved in phosphate buffer (20 ml; 0.1 mol/L; pH 7.0), was heated either under reflux or in a laboratory autoclave.

[b] The heating was performed in malonate buffer (0.1 mol/L, pH 7.0).

ble III, a higher temperature increased the amounts of Furaneol. Since already at boiling temperatures (expt. 1) substantial amounts of the odorant were formed, the data indicated that the precursors are relatively labile.

Free sugars as precursors. As mentioned in the introduction, rhamnose and fructose have been shown to liberate Furaneol by heat treatment. To reveal their role as precursors, aqueous solutions of rhamnose, fructose and, in addition, glucose and sucrose were heated at 100°C and 150°C and the amounts of Furaneol liberated were determined (Table IV). At 100°C only rhamnose yielded substantial amounts of Furaneol while fructose only yielded very small amounts. Glucose and sucrose were ineffec-

Table V. Concentrations of sugars and sugar phosphates
in the low molecular weight fraction from baker's yeast

Carbohydrate	Conc.(mg/kg yeast)[a]
Sucrose	2440
Fructose	33
Glucose	1265
Glucose-6-phosphate	125
Fructose-6-phosphate	230
Fructose-1,6-diphosphate	5750

[a] The concentrations were determined by enzymatic methods [17].

tive. At elevated temperatures the amounts liberated
from glucose and fructose were slightly enhanced, whereas the amount generated from rhamnose increased by a
factor of fourteen. Phosphate ions significantly favored
the Furaneol production. The amounts of Furaneol formed
from rhamnose decreased by a factor of more than forty,
when phosphate was replaced by malonate buffer (cf.
expts. 4 and 5). Additions of the amino acids proline or
alanine (0.2 mmol) did not enhance the formation of the
odorant from the carbohydrates (data not shown).

The free sugars present in the low molecular weight
fraction from yeast were analysed by thin layer electro-
phoresis [16] and quantified by enzymatic methods [17].
Three sugar phosphates were detected, among which fruc-
tose-1,6-diphosphate (FDP) predominated (Table V). Fructose was present only in low amounts, while rhamnose was
not detectable.

The amounts of Furaneol formed after thermal treatment of the three sugar phosphates were determined. The
data showed (Table VI) that the sugar phosphates are effective precursors of Furaneol but showing different reactivities. At 100°C only fructose-6- and fructose-1,6-
diphosphate liberated substantial amounts, while at
150°C also glucose-6-phosphate was active. Replacement
of the phosphate by malonate buffer (cf. expt. 4) revealed that substantial amounts of Furaneol were formed
from FDP also in the absence of phosphate ions. The data
revealed that FDP, the predominating free sugar in
yeast, is undoubtedly the most important precursor of
Furaneol. This is corroborated by the fact that the
amounts of Furaneol liberated from FDP in yeast or the
model experiment were well correlated.

Precursors in corn. An aqueous extract of corn flour was
fractionated by centrifugation and ultrafiltration. As
in yeast, the precursors of Furaneol could be located in

Table VI. Formation of Furaneol from sugar phosphates[a]

| No. | Sugar phosphate (0.2 mmol) | Furaneol (μg) | |
		100°C (60 min)	150°C (45 min)
1	Glucose-6-phosphate	0.6	29.7
2	Fructose-6-phosphate	16.4	51.4
3	Fructose-1,6-diphosphate	17.5	46.4
4	Fructose-1,6-diphosphate[b]	----	20.7

[a] The sugar phosphate was heated in phosphate buffer (20 ml; 0.1 mol/L; pH 7.0).

[b] The phosphate buffer was replaced by malonate buffer (pH 7.0).

Table VII. Formation of Furaneol from the fraction of water-soluble, low molecular weight compounds of corn flour

| Expt. | Heating conditions | | Furaneol (μg/kg corn flour) |
	time (min)	temp. (°C)	
1[a]	60	100	<10
2[b]	45	150	93
3[c]	15	170	1103

[a] The material (from 30 g corn flour), dissolved in phosphate buffer (20 ml; 0.1 mol/L; pH 7.0) was heated under reflux.

[b] cf. footnote a. Heating was performed in a laboratory autoclave.

[c] The material was intimately mixed with silica gel (3 g) and dry-heated.

the LMW-fraction. In a first experiment the influence of the heating conditions on the formation of Furaneol from the LMW-fraction of corn was examined. The data revealed (Table VII) that in contrast to the yeast experiments at 100°C or 150°C only low amounts of the flavor compounds were formed. On the other hand, the amounts increased by a factor of twelve when the flour extract was dry-heated. This process afforded nearly the same amount determined in popcorn (cf. Table I).

A quantification of the free sugars present in the corn extract revealed that sucrose predominated and relatively high amounts of glucose and fructose were present (Table VIII). Sugar phosphates were present in low amounts and rhamnose was absent. The sugars identified

Table VIII. Concentrations of carbohydrates in corn
flour[a]

Carbohydrate	Conc. (mg/kg corn)
Sucrose	2020
Fructose	880
Glucose	1190
Glucose-6-phosphate	77
Fructose-6-phosphate	25

[a] The amounts were determined by enzymatic methods [17].

Table IX. Formation of Furaneol from dry-heated carbohydrates[a]

Carbohydrate (0.2 mmol)	Furaneol (μg)
Sucrose	<0.1
Fructose	18.7
Glucose	11.9
Glucose-6-phosphate	8.4

[a] The carbohydrate was intimately mixed with silica gel
(3 g) and dry-heated for 15 min at 170°C.

in the corn extract were then dry-heated under conditions simulating the popping process. The results
revealed (Table IX) that substantial amounts of Furaneol
were formed from fructose and glucose. The amounts
liberated from glucose-6-phosphate were lower and
sucrose was ineffective. As glucose and fructose are the
predominating free sugars in corn, the data allow the
conclusion that these carbohydrates are the most
important precursors of Furaneol during the popping
process.

Formation of Furaneol from sugar phosphates. To explain
the formation of Furaneol from the sugar phosphates by
known sugar degradation reactions [15] a reduction reaction must be assumed. This prompted us to investigate
whether in the presence of a reducing compound like ascorbic acid the formation of Furaneol from FDP will be
favored.

As shown in Table X, additions of ascorbic acid enhanced the amounts of Furaneol liberated from FDP and
also fructose-6-phosphate by a factor of two. The
amounts of Furaneol produced from fructose were also increased, but remained low compared to the sugar phosphates.

Very recently, Ledl (personal communication) identified Furaneol as a reaction product from thermally de-

Table X. Influence of ascorbic acid on the formation of
Furaneol from carbohydrates

Carbohydrate (0.2 mmol)[a]	Furaneol (μg)	
	without	with[b]
	addition of ascorbic acid	
Fructose-6-phosphate	51.4	102.6
Fructose-1,6-diphosphate	46.4	90.7
Fructose	2.9	3.8

[a] The carbohydrate was dissolved in phosphate buffer (20 ml; 0.1 mol/L; pH 7.0) and heated for 45 min in an autoclave (150°C).

[b] Ascorbic acid (0.05 mmol) was added.

graded 2,5-dimethyl-2,4-dihydroxy-3(5H)-furanone (acetylformoine). To elucidate its role as precursor, acetylformoine (0.2 mmol) was degraded by heating for 60 min at 100°C in phosphate butter (20 mL; 0.1 mol/L; pH 7.0). The reaction afforded 176 μg of Furaneol, which was by a factor of 10 more than from FDP (cf. Table VI). This result indicated acetylformoine as important intermediate in the formation of Furaneol.

A reaction route from FDP to acetylformoine is proposed in Fig. 1. After a 2,3-enolization as a first step, the 1-phosphate group is eliminated leading to a 1-desoxyosone-6-phosphate. As the same intermediate will be formed from fructose-6-phosphate and as both sugar phosphates liberated nearly the same amounts of Furaneol (cf. Table VI), it can be concluded that the presence of the 1-phosphate group does not influence the reactivity of the sugar. Fructose was little effective as Furaneol precursor in aqueous solution (cf. Table IV). Therefore, it can be concluded that the key function in FDP is the phosphate group at carbon-6. As phosphate is a good leavening group, its elimination followed by enolization reactions will result in the acetylformoine.

The formation of Furaneol from acetylformoine can be explained as detailed in Fig. 2. The reactive intermediate is the open chain form of acetylformoine. After a reduction, which may occur either by a disproportionation or by a reaction with further enoloxo compounds like ascorbic acid, subsequent enolization and water elimination will directly afford the formation of Furaneol from this intermediate.

In a dry heating process, like popping of corn, glu-

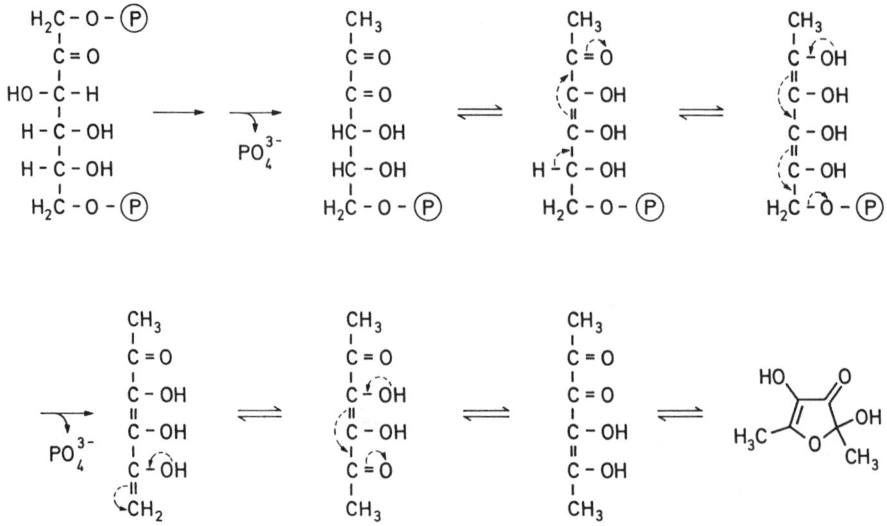

Figure 1. Formation of 2,5-Dimethyl- 2,4- dihydroxy-3(2H)- furanone (acetylformoine) from fructose-1,6-diphosphate (hypothesis)

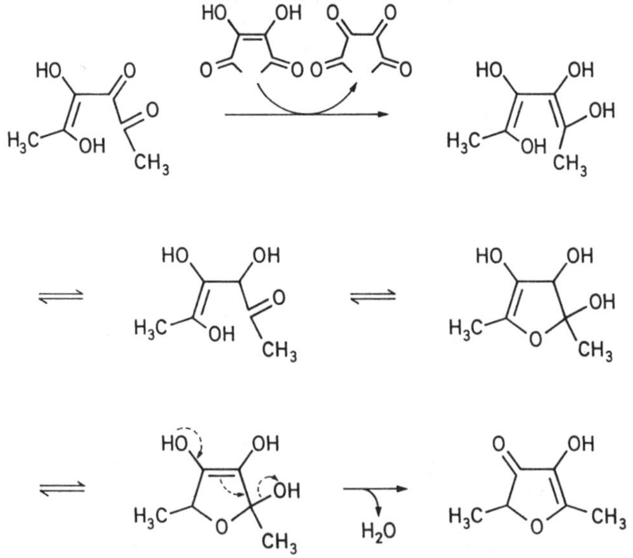

Figure 2. Formation of Furaneol from acetylformoine (hypothesis)

cose and fructose also can act as Furaneol precursors. Obviously under these conditions the elimination of the hydroxy group from carbon six is favored, thereby generating acetylformoine.

Conclusions

The data show that bakers yeast is the most important source of precursors for the caramel-like smelling odorant Furaneol in wheat bread crust. During the baking process, the thermally labile fructose-1,6-diphosphate, which was established as most important precursor in the yeast, is degraded to Furaneol via acetylformoine as intermediate.

Furaneol was also established as important odorant in fresh popcorn. Results of model experiments indicated that in the dry-heating process applied during popping of the corn the formation of Furaneol occurs by degradation of the less reactive non-phosphorylated sugars fructose and glucose, which predominated in the corn flour.

The data allow the conclusion that in heat-processed foods containing hexoses Furaneol will always be formed. Whether the amounts liberated are sufficient for a sensory contribution to the food flavor, strongly depends on the composition of the sugars present in the raw material and on the processing conditions.

Acknowledgements. I am grateful to Prof. F. Ledl (Institut for Food Chemistry, University of Stuttgart) for supplying a sample of acetylformoine and Mrs. D. Karrais and I. Liß for careful technical assistance. The work was financed by the AIF (Köln) and the Forschungskreis der Ernährungsindustrie (Hannover).

Literature cited

1. Pittet, A.O.; Rittersbacher, P.; Muralidhara, R.; *J. Agric.Food Chem.* **1970**, *18*, 929-933
2. Rodin, J.O.; Himel, C.M.; Silverstein, R.M. *J.Food Sci.* **1965**, *30*, 280-285
3. Willhalm, B.; Stoll, M.; Thomas, A.F. *Chem.Ind.* **1965**, 1629-1630
4. Mayerl, F.; Näf, R.; Thomas, A.F. *Phytochemistry* **1989**, *28*, 631-633
5. Tonsbeek, C.H.T.; Plancken, A.J.; v.d. Weerdhof, T. *J.Agric.Food Chem.* **1968**, *16*, 1016-1021
6. Takei, Y.; Yamanishi, T. *Agric.Biol.Chem.* **1974**, *38*, 2329-2336
7. Tressl, R.; Bahri, D.; Köppler, H.; Jensen, A. *Z. Lebensm.Unters.Forsch.* **1978**, *167*, 111 - 114
8. Schieberle, P. In *The Maillard Reaction in Food Processing, Human Nutrition and Physiology*; Finot,

P.A.; Aeschbacher, H.U.; Hurrell, R.F.; Liardon, R.;
Birkhäuser, Basel, **1990**, pp. 187-196

9. Schieberle, P. *J.Agric.Food Chem.* **1991**, *39*, 1141-1144

10. Hodge, J.E.; Fisher, B.E.; Nelson, E.C. *Am.Soc. Brewing Chem.Proc.* **1963**, 84

11. Shaw, P.E.; Berry, R.E. *J.Agric.Food Chem.* **1976**, *25*, 641-644

12. Doornbos, T.; v.d. Ouweland, A.M.; Tjan, S.B. *Progr. Food Nutr.Sci.* **1981**, *5*, 57-63

13. Mills, F.D.; Hodge, J.E. *Carbohydrate Research* **1976**, *51*, 9-21

14. Sen, A.; Schieberle, P.; Grosch, W. *Lebensm.Wiss. Technol.* **1991**, *24*, 364-369

15. Ledl, F. In *The Maillard Reaction in Food Processing, Human Nutrition and Physiology*; Finot, P.A.; Aeschbacher, H.U.; Hurrell, R.F.; Liardon, R.; Birkhäuser, Basel, **1990**, pp. 19 - 42

16. Scherz, H. *Z.Lebensm.Unters.Forsch.* **1985**, *181*, 40-44

17. Bergmeyer, H.U. *Methoden der enzymatischen Analyse*, Verlag Chemie, Weinheim, **1974**, pp. 1283-1287, 1359-

RECEIVED December 18, 1991

Chapter 13

Flavor Compounds Formed from Lipids by Heat Treatment

T. Shibamoto and H. Yeo

Department of Environmental Toxicology, University of California, Davis, CA 95616

Food components, including proteins, carbohydrates, amino acids, sugars, and lipids, degrade into smaller molecules which undergo secondary reactions to form many flavor chemicals in cooked foods. The best known reaction occurring during cooking foods is the Maillard reaction. Investigations of complex reactions occurring in heated foods most commonly use so-called Maillard model system consisting of a sugar and an amino acid. A sugar is used as a source of carbonyl compounds which react with amines from amino acid to produce flavor chemicals. Recently, lipids have begun to receive much attention as flavor precursors in cooked foods because lipids also produce many carbonyl compounds upon heat treatment. Aldehydes and ketones are precursors of many heterocyclic compounds which contribute roasted or toasted flavors to cooked foods.

Lipids, upon exposure to heat and oxygen, are known to decompose into secondary products including alcohols, aldehydes, ketones, carboxylic acids, and hydrocarbons. Aldehydes and ketones produce heterocyclic flavor compounds reacting with amines, such as amino acids, via Maillard-type reactions in cooked foods (1). One of the major food constituents, lipids have been known to produce many aldehydes and ketones, which may undergo secondary reaction with amine compounds to yield flavor chemicals in cooked foods. In fact, lipid-rich foods (2) or deep-fat-fried foods (3) reportedly produce many flavor chemicals. Therefore, lipids may provide the aldehydes and ketones essential to the flavor formation.

Volatile Compounds Formed in Beef Fat

Beef fat was heated with or without glycine in a pressurized bottle at 200 °C for 4 h and the volatile chemicals formed were isolated using a simultaneous steam

0097–6156/92/0490–0175$06.00/0
© 1992 American Chemical Society

distillation/solvent extraction apparatus (SDE)(*4,5*). The extract was analyzed by gas chromatography/mass spectrometry (GC/MS).   One hundred forty-three compounds were isolated and identified in the extracts.   The major compounds identified in the extract of beef fat alone were n-alkanes, n-alkenes, n-alcohols, n-aldehydes, n-alkylcyclohexanes, and 2-ketones. Various isoacid derivatives such as isoalcohols were also found in this sample. Proposed formation mechanisms of the major compounds identified in this study are shown in Figure 1. Table I shows the number of each chemical group isolated from heated beef fat.

Table I. Number of volatile compounds in different
chemical groups identified in heated beef fat

Compounds	Isolated by	
	SPE	SDE
n-Alkanes	7	15
n-Alkenes	31	12
n-Aldehydes	18	13
2-n-Ketones	6	13
n-Alcohols	4	12

Many nitrogen-containing compounds were isolated from the beef fat heated with glycine. This extract gave an unpleasant odor possibly due to the presence of certain aldehydes and ketones. Alkylpyridines were also detected in this extract. The amount of 2-butyl- and 2-pentylpyridine was found to decrease as the amount of glycine added increased. The formation of n-nonanal in large quantity (7.4%) in the beef fat samples and its consumption in the presence of glycine suggest n-nonanal as a possible precursor of 2-butylpyridine. Similarly, 1-decanal is a possible precursor of 2-pentylpyridine.

Homologues of 5-alkyldihydro-2(3H)-furanone were tentatively identified. As shown in Figure 1, the proposed formation mechanism of these compounds is from the RCOO• radical via ring closure between the carbonyl radical and γ-carbon atom followed by loss of γ-hydrogen as a radical. The formation of cyclohexane and alkylpyridine are well explained by RCOO•, RCH$_2$O•, RO•, and R• radicals formed from triglyceride.

The headspace volatiles from overheated beef fat were also isolated using a simultaneous purging extraction (SPE) apparatus and identified by GC/MS (*6*). The first column shows the numbers found in the extract from SPE and the second column shows the numbers identified in the extract from SDE. In the extract from SPE, 87 compounds were identified including 7 alkanes, 31 alkenes, 18 aldehydes, and 6 ketones. Aldehydes, which constituted 23.41% of the total volatiles isolated, were the major components identified in the extract from SPE.

Volatile Aldehydes Formed in Heated Pork Fat

Aldehydes and ketones formed in a headspace of heated pork fat were trapped in an aqueous cysteamine solution and recovered using an SPE (7). Table II shows aldehydes and ketones found in the extract.

Table II. Aldehydes and ketones determined in headspace of heated pork fat

Compounds	Concentration (mg/L)
Acetaldehyde	49.1
Formaldehyde	7.99
Propanal	71.3
2-Pentanone	1.25
Butanal	224
2-Hexanone	6.51
Pentanal	1003
$C_5H_{11}CHO$ (branched)	13.7
2-Heptanone	11.4
Hexanal	3090
$C_6H_{13}CHO$ (branched)	40.1
2-Octanone	15.2
Heptanal	1261
$C_7H_{15}CHO$ (branched)	423
$C_7H_{15}CHO$ (branched)	135
Octanal	251
$C_8H_{17}CHO$ (branched)	30.7
$C_8H_{17}CHO$ (branched)	13.2
Nonanal	80.8

Nine aldehydes and four ketones were identified. The production of formaldehyde, which has never been reported prior to this study, was determined in appreciable amounts in the pork fat samples. A series of straight chain-aldehydes (n-C_1 to n-C_9) were identified. Hexanal was one of the major aldehydes produced. This suggests that linoleic acid esters are a major constituent of pork fat because it is generally recognized that the oxidative cleavage of double bonds produces aldehydes or ketones (8). Other major aldehydes recovered such as heptanal, nonanal, and butanal may correspond to the amount of possible fatty acid precursors in pork fat.

Acrolein Formed in Heated Cooking Oils

Acrolein, the simplest α,β-unsaturated aldehyde, has been found in various foods such as cooked horse mackerel (9). Acrolein produced from cooking oil heated at various temperatures in headspace was analyzed as a morpholine derivative (10). Table III shows amounts of acrolein found in a headspace of various cooking oils.

Table III. Amount of acrolein recovered in headspace of 200 g of heated cooking oils using SPE with N_2 or Air stream

| Cooking oil | Iodine value | Amount (mg) | |
		N_2	Air
Corn oil	103 - 128	54.08	81.05
Soybean oil	120 - 141	29.55	76.11
Sunflower oil	125 - 136	36.90	57.61
Olive oil	80 - 88	72.90	103.63
Sesame oil	103 - 195	58.98	85.51

The amount of acrolein formed appear to be inversely proportional to the iodine values which indicate the degree of unsaturation of fatty acid (11). Olive oil, which has the lowest iodine value among the oils used, produced the most acrolein while soybean oil, which has the highest iodine value, produced the least acrolein. The major pathway of acrolein formation in heated fats or oils is dehydration of glycerol which is formed from hydrolysis of triglycerides (12).

Acrolein has been known as a lachrymator, and the vapor causes eye, nose, and throat irritation. However, an appropriate amount of acrolein stimulates the taste and olfactory organs and consequently plays an important role in physiological effects of cooked lipid-rich foods or deep-fat fried foods.

Volatile Compounds Formed from a Corn Oil/Glycine Maillard Model System

Headspace volatiles obtained from a corn oil heated with or without glycine were isolated using the SPE apparatus and identified by GC/MS (13). The major compounds identified in the corn oil samples were aldehydes, hydrocarbons, and ketones. With the addition of glycine, five of the unsaturated aldehydes formed in the corn oil samples were not detected. They were trans-2-butenal, trans-2-pentenal, cis-2-hexenal, cis-2-heptenal, and 2-nonenal. The presence of glycine did not influence the amount of hydrocarbons, alcohols, esters, and aromatic compounds formed.

Figure 2 shows the total amounts of unsaturated aldehydes formed with various amounts of glycine in corn oil. The addition of glycine to corn oil decreased the amount of volatile unsaturated aldehydes by almost 100 times, suggesting that secondary reactions occurred between glycine and the aldehydes. Figure 3 shows the relative amounts of nitrogen- containing compounds produced in various corn oil/glycine mixtures. The amount of these compounds

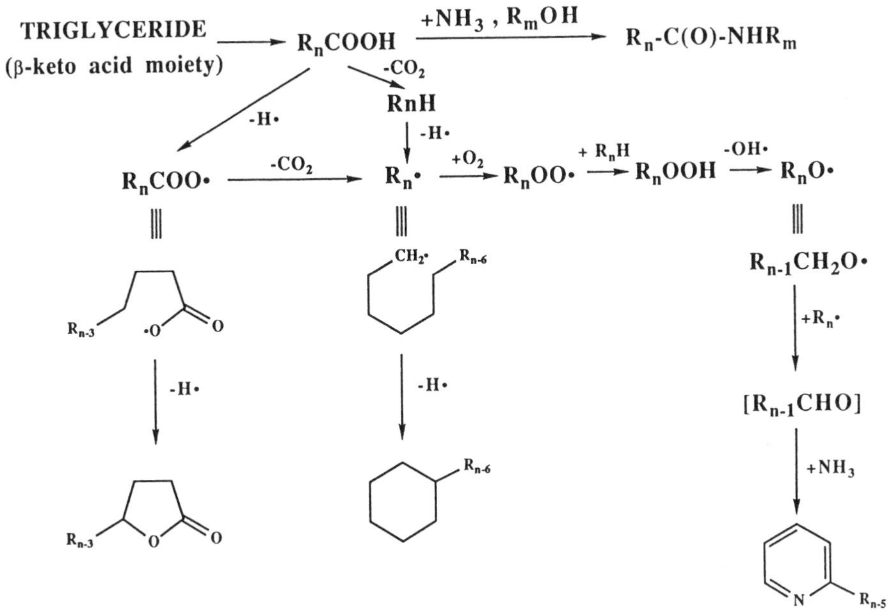

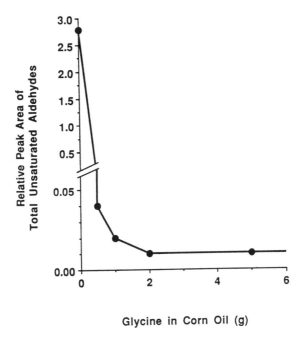

Figure 1. Proposed formation pathways of some major volatiles formed from lipids.

Figure 2. Total amounts of unsaturated aldehydes formed with various amounts of glycine in corn oil.

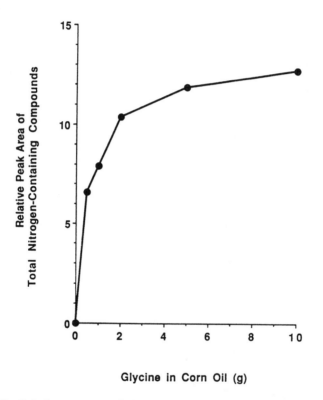

Figure 3. Relative amounts of nitrogen-containing compounds produced in various corn oil/glycine mixtures.

increases with increasing amounts of glycine in corn oil. This observation is mainly due to the formation of nitriles, pyridines, and pyrroles. The majority of the monoalkylpyridines were 2- or 3-substituted. 1-Methyl-2-propylpyrrole, 1-methyl-2-butyl-pyrrole, and 1-methyl-2-pentylpyrrole have never been reported prior to this study.

Volatile Compounds Formed in Heated Lipid-Rich Foods, Eggs

Eggs generally do not possess strong characteristics flavors. However, they contain amino acids, proteins, carbohydrates and lipids in large quantities, suggesting the possibility of flavor formation via the Maillard reaction. Headspace volatiles from heated whole egg, egg white, and egg yolk were isolated using an SPE apparatus and identified by GC/MS (*14*). A total of 141 volatile compounds were identified in cooked whole egg, egg yolk, and egg white. The percent yields of total volatiles from cooked whole egg, egg yolk, and egg white were 0.0083, 0.0030, and 0.0014%, respectively. The amount total volatiles produced from whole egg is considerably greater than did either egg yolk or egg white alone. This may be due to the presence of many precursors of the Maillard reaction products in whole egg. Table IV shows major volatile compounds found in heated eggs.

Aldehydes were the major volatiles in cooked egg yolk. Whole egg and egg white, however, yielded only a few volatile aldehydes. 2-Methylpropanal, a major aldehyde in egg yolk and whole egg volatiles, may be formed from Strecker degradation of valine or oxidative decomposition of lipids (*15*), as egg yolk is rich in both valine and lipids (*16*). Low-molecular-weight ketones such as acetone were found in large amounts in whole egg and egg white, whereas higher molecular weight 2-methylketones (C_7-C_{10}) were found only in egg yolk. Several alcohols, sulfides which posses a characteristic cooked egg flavor, and a large number of nitriles were found in the three egg samples.

Table IV.   Major volatile compounds identified in cooked egg samples

	Total GC peak area%		
Compounds	Whole egg	Egg yolk	Egg white
Aldehydes	10.73	64.12	4.53
Ketones	12.43	4.16	31.08
Nitriles	20.73	0.04	17.61
Pyrazines	6.65	8.69	18.24
Pyrroles	6.64	2.27	3.22
Pyridines	2.81	a	1.10
Alkylbenzenes	20.07	1.89	4.66
Furans	0.52	2.47	0.69
Thiazoles	a	0.18	1.83

a Not detected

Nitrogen-containing heterocyclic compounds such as pyridines, pyrazines, pyrroles, and thiazoles were the major flavor compounds identified in the egg samples. Many alkylpyrazines identified in the present study are known as the products of the Maillard reactions. They contribute roasted or toasted flavors to cooked foods. Some pyrroles reportedly contributed off-flavors to certain cooked food products. However, 2-acetylpyrrole, found predominantly in whole egg sample, possess a pleasant, caramel-like flavor. Egg white, which is rich in proteins, produced thiazoles in large quantities. Thiazoles are proposed to form from sugar and sulfur-containing amino acid.

As these reports show, lipid must play an important role in the formation of flavor chemicals in cooked foods, in particular in lipid-rich foods.

Literature Cited

1. Shibamoto, T. *Instrumental analysis of foods.* Vol. I. G. Charalambous and G. Inglett Eds. Academic Press, New York, **1983**, 229.
2. Mottram, D. S. Edwards, R. A. *J. Sci. Food Agric.* **1983**, *34*, 517.
3. Ho, C. T. Carlin, J. T. Huang, T. C. *Flavor Science and Technology,* Martens, M. Dalen, G. A. Russwum, H. Eds, Wiley & Sons, London, **1987**, 35.
4. Ohnishi, S. Shibamoto, T. *J. Agric. Food Chem.* **1984**, *32*, 987.
5. Likens, S. T. Nickerson, G. B. *J. Chromatogr.* **1966**, *21*, 1.
6. Umano, K. Shibamoto, T. *J. Agric. Food Chem.* **1987**, *35*, 14.
7. Yasuhara, A. Shibamoto, T. *J. Food Sci.* **1989**, *54*, 1471.
8. Mottram, D. S. Edwards, R. A. MacFie, H. J. H. *J. Sci. Food Agric.* **1982**, *33*, 934.
9. Shinomura, M. Yoshimatsu, F. Matsumoto, F. *Kaseigaku Zasshi* **1971**, *22*, 106.
10. Umano, K., Shibamoto, T. *J. Agric. Food Chem.* **1987**, *35*, 909.
11. Codd, L. W. Dijkhoff, K. Fearon, J. H. van Oss, C.J. Roebersen, H. G. Stanford, E. G., Eds. "Oils and Fats". In *Chemical Technology, an Encyclopedic Treatment,* Barnes & Noble: New York, **1975**, Vol 8.
12. Adkins, H. Hartung, W. H. "Acrolein". In *Synthesis Organiques*; Masson: Paris, 1935.
13. Macku, C. Shibamoto, T. *J. Agric. Food Chem.* **1991**, *39*, 1265.
13. Umano, K. Hagi, Y. Shoji, A. Shibamoto, T. *J. Agric. Food Chem.* **1990**, *38*, 461.
14. Beltz, H. D. Grosch, W. Eggs. In *Food Chemistry*; Springer-Verlag: Berlin, **1987**.
15. Lundberg, W. O. *Autoxidation and Autoxidants*; Interscience: New York, **1962** Vol. I.

RECEIVED December 18, 1991

Chapter 14

Formation of Meatlike Flavor Compounds

Werner Grosch and Gabriele Zeiler-Hilgart

Deutsche Forschungsanstalt für Lebensmittelchemie, Lichtenbergstrasse 4, D–8046 Garching, Germany

The formation of 2-methyl-3-furanthiol (MF-SH), its oxidation product bis(2-methyl-3-furyl)disulfide (MF-S$_2$) and 2-furfurylthiol (FF-SH) were determined in model systems containing the precursors in concentration levels approximated to those occurring in meat. The degradation of thiamine was more effective in the production of MF-SH than the reaction of ribose and cysteine. Addition of cysteine, and, to a lesser extent, of H$_2$S enhanced the formation of MF-SH from thiamine. In contrast to thiamine its pyrophosphate derivative was inactive as precursor of MF-SH. The amounts of FF-SH and MF-S$_2$ formed were very low in all models investigated.

2-Methyl-3-furanthiol (MF-SH), its oxidation product bis(2-methyl-3-furyl)disulfide (MF-S$_2$) and 2-furfurylthiol (FF-SH) have been identified as impact compounds of the "meaty" odor note of boiled beef, pork and chicken meat (1-3).

Model experiments have indicated that MF-SH and MF-S$_2$ are formed by a degradation of thiamine (4, 5), by a Maillard-type reaction of ribose and cysteine (6, 7) and recently by a reaction of inosine 5'-monophosphate with cysteine (8). The latter two reactions have also been established as source of FF-SH (6-8).

4-Hydroxy-5-methyl-3(2H)-furanone (from ribose) and hydrogen sulphide (from cysteine) have been suggested (6, 9) as intermediates in the Maillard reaction to MF-SH. FF-SH is probably formed by the reaction of H$_2$S with the sugar breakdown product furfural (6, 10).

0097–6156/92/0490–0183$06.00/0
© 1992 American Chemical Society

Table I. Compounds Discussed as Precursors of MF-SH, FF-SH and MF-S$_2$: Concentrations in Raw Beef and Pork (13-18)

Compound	Beef	Pork
	(mg/kg wet weight)	
Ribose	10-1300	-[a]
Total SH	690 (2530)[b]	627 (2300)[b]
Nonprotein SH[c]	11 (42)[b]	5 (19)[b]
GSH	71	194
Thiamine	0.9-2.3	8-10

[a] Data not available.

[b] Values in brackets: SH-content expressed as cysteine.

[c] Without GSH.

During boiling of meat, H_2S is released by the degradation of cysteine occurring free and bound, e.g. in reduced glutathione (GSH) and muscle proteins (11, 12). Table I contains data about the concentration levels in both, raw beef and pork, of free and bound cysteine, GSH, thiamine and ribose.

In the following study the formation of MF-SH, MF-S$_2$ and FF-SH was determined in model systems containing their precursors in concentration levels approximated to those occurring in raw beef and pork meats (Table I). In the models, the pH value was adjusted to 5.7, as meat is characterized by a pH ranging between 5.5 and 6.0 (19). A boiling period of 2 h which is usual for the preparation of a broth was chosen.

The aim of the present study was to estimate the actual contribution of each of the precursors discussed to the formation of MF-SH, MF-S$_2$ and FF-SH under the conditions used for the boiling of meat.

Experimental

The components of the reaction systems detailed in Tables II-VI, and VIII were dissolved in Na-pyrophosphate buffer (pH 5.7, 0.2 mol/L, 500 mL) unless otherwise specified. Two different methods were used to boil the reaction systems and to remove the volatile compounds.

Method I: the mixture was refluxed for 2 h and extracted after cooling. Method II: the mixture was distilled and continuously extracted (SDE) for 2 h in the apparatus according to Nickerson and Likens.

MF-SH, FF-SH and MF-S$_2$ were quantified in the cooled reaction mixture (Method I) and in the SDE extract (Method II) by mass chromatography using the corresponding deuterated flavor compounds (d-MF-SH, d-FF-SH, d-MF-S$_2$) as internal standards (20, 21).

Table II. Thiamine as Precursor

No.	Reaction system[a]	N[b]	Amount (μg)[c] MF-SH	FF-SH	MF-S$_2$
1	Thiamine	3	4.7±1.2	1.9±0.4	<0.2
2	Thiamine pyrophosphate	2	<0.5	<0.4	<0.2
3	Thiamine + ribose	2	5.2±0.4	<1.5	n.a.
4	Thiamine + cysteine	4	33±3.7	<1.5	n.d.
5	Thiamine pyrophosphate + cysteine	2	<0.5	<0.5	<0.2
6	Thiamine + cysteine + ribose	3	19±2.5	1.2±1.1	<0.2
7	Thiamine + GSH + ribose	3	23±1.2	<1.7	<0.2
8	Thiamine + GSH + cysteine + ribose	4	9.0±2.8	2.3±1.0	1.6±1.2

[a] Reactants: thiamine · HCl (10.2 mg; 30 μmol), thiamine pyrophosphate (13.9 mg; 30 μmol), D(-)-ribose (100 mg; 670 μmol), cysteine · HCl (57 mg; 325 μmol), GSH (71 mg; 230 μmol).

[b] N: Number of trials.

[c] Amount (mean ± standard deviation) found after isolation by method I.

n.a.: not analysed; n.d.: not detectable.

Results

Thiamine as precursor. In the model experiments listed in Table II thiamine alone as well as in combinations with ribose, cysteine and GSH were studied as precursor of MFSH, FF-SH and MF-S$_2$.

Experiment no. 1 indicates that the boiling of 30 μmol of thiamine for 2 h resulted in the formation of 41 nmol of MF-SH and 17 nmol of FF-SH. In contrast to thiamine, its pyrophosphate derivative was inactive as a precursor of the three sulfur compounds (no. 2). Addition of ribose to thiamine (no. 3) did not significantly change the amount of MF-SH formed. Cysteine, however, was very effective (no. 4): the yield in MF-SH increased 7-fold compared to the amount of MF-SH obtained in experiment no. 1 (thiamine without additions), while the amount of FF-SH formed was not enhanced. Also in the presence of cysteine, thiamine pyrophosphate was inactive as precursor of MF-SH (no. 5). Addition of ribose to the model thiamine/cysteine reduced the production of MF-SH from 33 to 19 μg (nos. 4 and 6). Replacement of cysteine in the mixture no. 6 by a lower concentration of GSH (230 μmol instead of 325 μmol) increased a little the yield of MF-SH (no. 7), while the complete reaction system consisting of thiamine, GSH, cysteine and ribose

Table III. Influence of the Levels of Cysteine and Thia-
           mine on the Formation of MF-SH in the Reac-
           tion System Thiamine/Ribose/Cysteine

| Cysteine (μmol)[a] | Thiamine[a] | |
| | 30 μmol | 6 μmol |
	MF-SH (μg)	
0	5.2	n.a.
85	14	n.a.
170	18	2.1
255	18	n.a.
325	19	n.a.
340	14	2.8
680	10	1.6
1360	4.8	0.5
2720	2.0	n.a.

[a] The reactants were dissolved in Na-pyrophosphate buffer (pH 5.7, 0.2 mol/L, 500 mL) containing D(-)-ribose
(100 mg; 670 μmol).

n.a.: not analysed

(no. 8) produced only 27 % of MF-SH found after the
boiling of thiamine/cysteine (no. 4).

The latter result suggested that the concentration of
cysteine (free or bound) affects the amount of MF-SH
formed. To prove this suggestion, model systems containing thiamine in two concentrations, a constant amount of
ribose and varying amounts of cysteine were studied.
Table III shows that the higher concentration of thiamine yielded throughout higher amounts of MF-SH. The influence of cysteine was also very strong as, in a concentration range of 170 to 325 μmol it produced the maximum amount of MF-SH, while a higher concentration was
inhibitory.

Also pyrophosphate affected the formation of MF-SH by
the model thiamine/ribose/cysteine (Table IV). Only 35 %
to 51 % was found, when the pyrophosphate concentration
was halved.

Studies on the system ribose/cysteine. A comparison of
the model systems nos. 4 and 6 in Table II suggests that
the combination of ribose and cysteine is not very effective in the production of MF-SH, as the addition of
ribose to the mixture of thiamine and cysteine decreased
the yield of MF-SH.

To clarify this point, the mixture of ribose and cysteine was heated under different conditions (Table V).
The first model system was the same as reported under
no. 6 in Table II, but the thiamine was omitted. The results (Table V) indicate that only traces of MF-SH were
detectable after the boiling of the reactants for 2 h.

Table IV. Influence of the Levels of Pyrophosphate and
          Cysteine on the Formation of MF-SH in the Re-
          action System Thiamine/Ribose/Cysteine

| Cysteine (μmol)[a] | Pyrophosphate (mol/L)[a] | |
| | 0.2 | 0.1 |
	MF-SH (μg)[b]	
325	19	8.8
340	14	7.2
680	10	3.5

[a] The cysteine was dissolved in Na-pyrophosphate buffer
(pH 5.7, 0.1 or 0.2 mol/L, 500 mL) containing D(-)-ri-
bose (100 mg; 670 μmol) and thiamine. HCl (10.2 mg;
30 μmol).
[b] Isolated by method I.

Table V. Reaction System Cysteine/Ribose

No.	Reaction system D(-)-Ribose Cysteine (μmol $\cdot$ mL^{-1})		Amounts (μg)[a] MF-SH	FF-SH	MF-S$_2$
1[b]	1.34	0.64	<0.1	n.d.	n.d.
2[c]	30	41	0.6	0.4	n.d.

[a] Isolated by method I; mean values of 3 (no. 1) and 2
(no. 2) trials.
[b] The compounds dissolved in Na-pyrophosphate buffer (pH
5.7, 0.2 mol/L, 500 mL) were refluxed for 2 h.
[c] The compounds dissolved in Na-phosphate buffer (pH
5.6, 0.5 mol/L, 2 mL) were heated for 1 h at 140°C according to *Farmer* and *Mottram* (7).

n.d.: not detectable

    As previously mentioned, *Farmer and Mottram* (7) have
studied the model cysteine/ribose. However, they used
much higher concentrations of both reactants in a phosphate buffer of pH 5.6 and, also, they heated the mixture for 1 hour at 140°C. The authors recognized relatively big peaks of MF-SH and FF-SH in the gas chromatogram of the volatile products, but did not give exact
figures about the amounts formed.
    We have repeated the experiment reported by *Farmer*
and *Mottram* (7). MF-SH and FF-SH were detectable, but
only in quite low concentrations (no. 2 in Table V).

Table VI. Influence of Hydrogen Sulphide on the Formation of MF-SH, FF-SH and MF-S$_2$ from Different
Compounds

No.	Compound[a]	Amount (μg)[b] MF-SH
1	Thiamine	15 ± 7[c]
2	Ribose	<0.2
3	Ribose-5-phosphate	0.4
4	Ribulose	<0.2
5	5'-Inosinic acid	0.2

[a] Thiamine · HCl (30 μmol) and the other compounds(670
μmol each) were dissolved in the pyrophosphate buffer.
Before boiling and during the 30th and 90th min of
boiling the solution was flushed with H$_2$S (200 mL
each); the total amount of H$_2$S was 26.8 mmol.

[b] Isolated by method I.

[c] Mean ± standard deviation of four determinations.

Beef is relatively high in 5'-ribomononucleotides
(22) which might release ribose during boiling. Therefore, inosinic acid (1.3 g) as an example for a mono-
nucleotide and cysteine (2.5 g), dissolved in the pyrophosphate buffer of pH 5.7 (500 mL), were tested as
source of MF-SH. However, this mixture was also not pro-
ductive, as only traces were formed during the boiling
period of 2 h (data not shown).

Hydrogen sulphide as precursor. It has been suggested
(6, 9) that H$_2$S released from cysteine during the heat
treatment of meat is involved in the formation of MF-SH.
To check this suggestion, the precursors summarized in
Table VI were treated 3 times with H$_2$S during the boiling process. The total amount of H$_2$S used in these experiments was calculated on the basis that all cysteine
occurring in beef (2530 mg/kg, Table I) was degraded to
H$_2$S.
Table VI shows that only thiamine yielded a substantial amount of MF-SH, thus, in an amount nearly three
times as high as in the experiment without H$_2$S (no. 1 in
Table II). However, as reported above (no. 4 in Table
II), addition of cysteine to thiamine resulted in a much
higher formation of MF-SH than the addition of H$_2$S. Ribose and the other compounds listed in Table VI were not
active in producing significant amounts of MF-SH in the
presence of H$_2$S.
Boiling of meat samples. The production of the sulfur
compounds by different samples of pork and beef meat was
compared. As the amounts of FF-SH and MF-S$_2$ lay below
the detection limits of 1.0 μg and 0.2 μg, respectively,
only MF-SH was assayed. Table VII indicated the forma-

Table VII. Formation of MF-SH during the Boiling of Meat

Meat	Sample	MF-SH (μg)[a]
Pork, shoulder without skin	A	1.0
	B	1.6
	C	1.9
Beef, top round (inside)	D	<0.2
	E	0.8
Beef, eye of round	F	2.1
Beef, rib wing	G	2.1
	H	2.1

[a] Meat (500 g) was minced and then suspended in tap water (500 mL). Boiling and isolation of the volatiles was performed by SDE (Method II).

tion of 1.0 to 1.9 μg of MF-SH during the boiling of 500 g of pork for 2 h. In the case of beef meat, boiling of the rib wing, which is used for the preparation of a bouillon, yielded relative high amounts of MF-SH.

Quantification of MF-SH was successful only, when the volatiles formed were continuously removed from the reaction system by SDE (method II). To demonstrate the differences between methods, the model systems in Table VIII were analysed by both methods. In the cases of thiamine (no. 1) and thiamine in combination with cysteine (no. 2), MF-SH increased approximately 5- and 4-fold, respectively, when it was isolated by SDE. In contrast, the amount of MF-SH resulting from ribose/cysteine was also very low in the extract obtained by method II.

Discussion

The results show that, at the low concentration levels occurring in meat, the degradation of thiamine was more effective in the production of MF-SH than the reaction of ribose and cysteine. Furthermore, addition of cysteine, and to a lesser extent H_2S, enhanced the formation of MF-SH from thiamine. In contrast, the amounts of FF-SH and $MF-S_2$ were always very low after the boiling of the model systems.

The oxidation of MF-SH to $MF-S_2$, taking place during the storage of the extracts before HRGC/MS-analysis, did not influence the results, since the addition of the deuterated internal standards immediately after the boiling period would compensate alterations caused by this reaction.

<u>Table VIII.</u> Comparison of Two Methods Used for the Isolation of MF-SH

No.	Reaction system	Method I[a]	Method II[a]
		MF-SH (μg)[b]	
1	Thiamine (30 μmol)	4.7 ± 1.2	25 ± 5
2	Thiamine (30 μmol) + cysteine (325 μmol)	33 ± 3.7	136 ± 0
3	Ribose (670 μmol) + cysteine (325 μmol)	<0.1	0.1

[a] The reaction mixture was either refluxed for 2 h and extracted after cooling (method I) or it was distilled and continuously extracted for 2 h (method II).

[b] Mean ± standard deviation of at least two determinations.

Van der Linde et al. (*4*) have heated a buffered (phosphate) solution of thiamine at 130°C and have identified numerous volatiles. To explain the formation of these compounds, they suggested that thiamine was hydrolysed by the heat treatment with formation of 5-hydroxy-3-mercaptopentan-2-one, the key intermediate (1 in Figure 1) in the reaction pathway to MF-SH (5 in Figure 1).

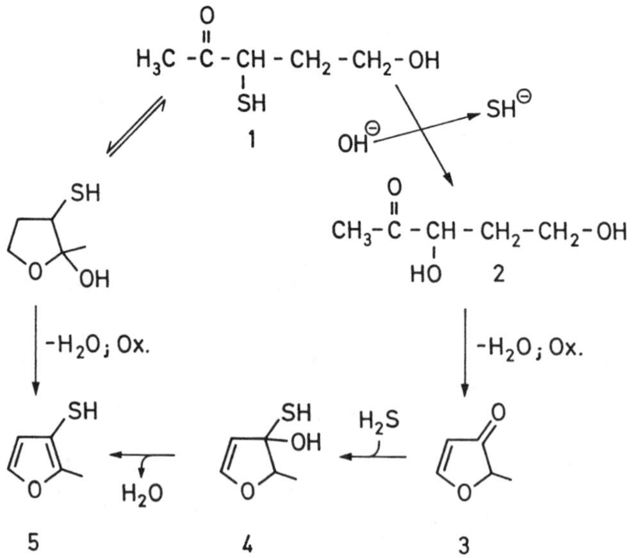

Figure 1. Hypothetical reaction route to explain the increase in 2-methyl-3-furanthiol, when thiamine is heated in the presence of H_2S.

The authors (4) assumed that a second reaction sequence, which started with the replacement of the SH-group in thiol 1 by a HO-group, competed with the formation of MF-SH (Figure 1). We suggest that this competition is influenced by the H_2S released from cysteine. As shown in Figure 1, H_2S may add to the oxofuran 3 with formation of the intermediate 4, which yields MF-SH by dehydration. This reaction sequence can explain the enhancement of MF-SH, when thiamine is heated in the presence of cysteine or H_2S. Another possiblity is that higher concentrations of H_2S may shift the equilibrium of SH-group replacement thus stabilizing compound 1 and increasing MF-SH formation.

The results reported here indicate, in addition, that only low amounts of MF-SH are produced during the boiling of meat. A more detailed investigation should clarify the factors influencing the production of MF-SH in meat.

Literature cited

1. Gasser, U.; Grosch, W. *Z.Lebensm.Unters.Forsch.* **1988**, *186*, 489-494.
2. Gasser, U.; Grosch, W. *Z.Lebensm.Unters.Forsch.* **1990**, *190*, 3-8.
3. Gasser, U.; Grosch, W. *Lebensmittelchemie* **1991**, *45*, 15.
4. Van der Linde, L.M.; van Dort, J.M.; de Valois, P.; Boelens, H.; de Rijke, D. In *Progress in Flavour Research*; Land, D.G.; Nursten, H.G., Eds.; Applied Sciences: London, **1979**; pp. 219-224.
5. Hartman, G.J.; Carlin, J.T.; Scheide, J.D.; Ho, C.-T. *J.Agric.Food Chem.* **1984**, *32*, 1015-1018.
6. Farmer, L.J.; Mottram, D.S.; Whitfield, F.B. *J.Sci. Food Agric.* **1989**, *49*, 347-368.
7. Farmer, L.J.; Mottram, D.S. *J.Sci.Food Agric.* **1990**, *53*, 505-525.
8. Zhang, Y.; Ho, C.-T. *J.Agric.Food Chem.* **1991**, *39*, 1145-1148
9. Van den Ouweland, G.; Peer, H.G. *J.Agric.Food Chem.* **1975**, *23*, 505-505.
10. Shibamoto, T. *J.Agric.Food Chem.* **1977**, *25*, 206-208.
11. Mecchi, E.P.; Pippen, E.L.; Lineweaver, H. *J.Food Sci.* **1964**, *29*, 393-399.
12. Ohloff, G.; Flament, I.; Pickenhagen, W. *Food Rev. Int.* **1985**, *1*, 99-148.
13. Sulser, H.: *Die Extraktstoffe des Fleisches.* Wissenschaftliche Verlagsgesellschaft, Stuttgart, **1978**, p. 22.
14. Gazzani, G.; Cuzzoni, M. *Riv.Soc.Ital.Sci.Aliment.* **1985**, *14*, 369-372.
15. Seuß, I.; Martin, M.; Honickel, K.O. *Fleischwirtsch.* **1990**, *70*, 913-919.
16. Hofmann, K.; Hamm, R. *Adv.Food Res.* **1978**, *24*, 1-111.
17. Wierzbicka, G.T.; Hagen, T.M.; Jones, D.P. *J.Food Compos.Anal.* **1989**, *2*, 327-337.

18. Scherz, H.; Senser, F. *Food Composition and Nutrition Tables 1989/90.* Wissenschaftliche Verlagsgesellschaft, Stuttgart, **1989**, pp. 207, 224.
19. Lawrie, R.A. *Meat Science*, 4th Ed., Pergamon Press, **1985**, pp. 51-53.
20. Grosch, W.; Sen, A.; Guth, H.; Zeiler-Hilgart, G. In *Flavour Science and Technology*; Bessière, Y.; Thomas, A.F.; Eds.; John Wiley & Sons: Chichester, **1990**, pp. 191-194.
21. Sen, A.; Grosch, W. *Z.Lebensm.Unters.Forsch.* **1991**, *192*, 541-547.
22. Baines, D.A; Mlotkiewicz, J.A. In *Recent Advances in the Chemistry of Meat*; Bailey, A.J., Ed.; The Royal Society of Chemistry: London, **1984**; pp. 119-164.

RECEIVED December 18, 1991

Chapter 15

Peptides as Flavor Precursors in Model Maillard Reactions

Chi-Tang Ho[1], Yu-Chiang Oh[1], Yuangang Zhang[1], and Chi-Kuen Shu[2]

[1]Department of Food Science, Rutgers University—The State University of New Jersey, New Brunswick, NJ 08903
[2]Bowman Gray Technical Center, R. J. Reynolds Tobacco Company, Winston-Salem, NC 27102

Equimolar aqueous solutions of glycine, diglycine, triglycine and tetraglycine were heated separately with D-glucose at 180°C at pH 4-5 in a Hoke sample cylinder for 2 hr. The Maillard reactions of glucose with glycine and triglycine produced a significantly greater amount of pyrazines than either diglycine or tetraglycine. The similarity of the results of glycine with triglycine and diglycine with tetraglycine in the pyrazine formation also suggests that tripeptides or tetrapeptides could be degraded through diketopiperazines. Alkyl 2(1H)-pyrazinones were identified as peptide-specific Maillard reaction products. The volatile products generated from the Maillard reaction of glucose with glutathione and its constituent amino acids in an aqueous medium were also compared.

The Maillard reaction is a well-known reaction that occurs in food during cooking. Because of the complexities of the Maillard reaction, many investigations have been aimed at understanding the mechanisms of the reaction. The pathways for the Maillard reaction orginally proposed by Hodge (1) have gained wide acceptance.

Recent works on the generation of aroma compounds from the Maillard reaction were mostly concerned with model systems using amino acids. Some of the amino acids used in model systems were proline (2-5), hydroxyproline (6), serine and threonine (7), cysteine (8-9), leucine (10) and glycine (11-12).

Although a wide range of peptides has been reported in considerable quantity in many food systems such as aged sake (13), meat (14) and hydrolyzed vegetable protein (15), the role of peptides as precursors in the generation of flavor compounds has not been investigated to an appreciable extent. Chuyen et al. (16) studied the reaction of various depeptides with glyoxal and reported the identification of 2-(3'-alkyl-2'-oxopyrazin-1')alkyl acids as major

0097–6156/92/0490–0193$06.00/0
© 1992 American Chemical Society

products. Most recently, Rizzi (17) reported that model Maillard
reactions of dipeptides and tripeptides with fructose generated
Strecker degradation products, such as Strecker aldehydes and alkylpyrazines, from amino acids with blocked amino and carboxyl func-
tionalities.

Volatile Compounds Formed from Maillard Reaction of Glucose with Gly, Gly-Gly, Gly-Gly-Gly and Gly-Gly-Gly-Gly

Equimolar aqueous solutions of glycine, diglycine, triglycine and
tetraglycine were heated separately with D-glucose at 180°C at pH
4-5 in a Parr bomb for 2 hr. Each reaction mixture was adjusted to
pH > 12 with NaOH, then extracted with methylene chloride, containing an internal standard, in a separatory funnel by multiple extraction method (5X50 ml). The methylene chloride extracts were
dried over anhydrous sodium sulfate and concentrated by blowing with
nitrogen gas to a final volume of 0.2 mL. The volatile compounds
were then analyzed by gas chromatography and gas chromatography-mass
spectrometry as described previously (18).
    Table I lists the volatile compounds generated in these systems.
From the quantitative data, it was observed that glycine or triglycine generated a larger amount of alkylpyrazines than either diglycine or tetraglycine.
    It is also interesting to note that furfural and 5-(hydroxy-
methyl)furfural were produced in a greater quantity in the reaction
of diglycine and tetraglycine with glucose, as compared to the glycine or triglycine. 2-Acetylpyrrole and 2-formyl-5-methyl-pyrrole
were identified as trace components in these model reactions.
    The formation of diketopiperazines from the thermal degradation
of dried polyglycine has been reported by Hayase et al. (19). From
Table I, the relative abundance of pyrazines formed from glycine and
triglycine are very close. However, the amount of pyrazines formed
from diglycine or tetraglycine was considerably less as compared to
either glycine or triglycine. The triglycine could be degraded into
glycine and diglycine through diketopiperazine. While tetraglycine
was degraded primarily into diglycine, further degradation of diglycine into glycine could require more energy. On the other hand, the
degradation of peptides by direct hydrolysis without the intermediate formation of diketopiperazine cannot be ruled out.
    The reactivity order, tetraglycine > triglycine > diglycine >
glycine for the color formation in the browning reaction reported by
Chuyen et al. (16) differs from our observation for pyrazine formation. This clearly indicates that for the Maillard reaction the
melanoidin formation is different from aroma formation in mechanism
and reactivity.

Pyrazinones as Peptide-specific Maillard Reaction Products

As shown in Table I, three novel pyrazinones were identified in the
reaction of glucose with diglycine, triglycine or tetraglycine. The
pyrazinones identified were 1,6-dimethyl-2(1H)-pyrazinone, 1,5-di-
methyl-2(1H)-pyrazinone and 1,5,6-trimethyl-2(1H)-pyrazinone. The
mass spectrum of 1,6-dimethyl-2(1H)-pyrazinone is shown in Figure 1.

Table I. Amount of Volatile Compounds Generated by
Glycine, Diglycine, Triglycine and Tetraglycine
with Glucose at 180°C for 2 Hours

Compounds	$I_k{}^a$	Gly	di-Gly	tri-Gly	tetra-Gly
		mg/mole amino compound			
pyrazine	738	12.18	-	-	-
methylpyrazine	798	257.80	-	186.81	3.17
furfural	808	14.14	100.44	3.87	16.93
2,5-dimethyl-pyrazine	887	360.86	10.28	266.57	46.88
2,6-dimethyl-pyrazine	894	198.76	-	139.50	17.41
trimethylpyra-zine	980	486.77	8.11	375.98	54.52
2-acetylpyrrole	1058	1.36	-	1.59	-
tetramethyl-pyrazine	1065	71.44	-	55.63	-
5-(hydroxyl-methyl)-furfural	1208	125.61	797.04	-	127.41
1,6-dimethyl-2(1H)-pyra-zinone	1315	-	323.00	119.35	422.35
1,5-dimethyl-2(1H)-pyra-zinone	1379	-	t	t	1.37
1,5,6-trimethyl-2(1H)-pyra-zinone	1476	-	t	2.85	0.57

a Linear retention indices calculated according to Majlat et al.
(1974) on an HP-1 column.

* t = trace

The structure of the pyrazinones were confirmed by comparing their mass spectra and GC retention times with the authentic compounds synthesized by the reaction of diglycine with pyruvaldehyde. These three pyrazinones were formed by the reaction of diglycine with pyruvaldehyde in yields of 25.56%, 9.39% and 20.16% for 1,6-dimethyl-2(1H)-pyrazinone, 1,5-dimethyl-2(1H)-pyrazinone and 1,5,6-trimethyl-2(1H)-pyrazinone, respectively. The total pyrazinones amounted to 55.11%. According to the mechanism proposed by Chuyen et al. (16), various dipeptides would react with α-dicarbonyl compounds to yield the pyrazinone derivatives, 2-(3'-alkyl-2'-oxopyrazin-1'-yl)alkyl acids. On the other hand, the degradation of glucose would produce various α-dicarbonyl compounds such as glyoxal, pyruvaldehyde and diacetyl. As shown in Figure 2, in the case of pyruvaldehyde, the amino end of the dipeptide would prefer to react with the aldehydic carbonyl group than the other ketone carbonyl carbon because of a steric effect. After Amadori rearrangement, the intermediate of dipeptide-pyruvaldehyde was cyclisized to form 2-(3'-alkyl-2'-oxopyrazin-1'-yl)alkyl acids. At the elevated temperature (180°C) used in our model experiment, the 2-(3'alkyl-2'-oxopyrazin-1'-yl)alkyl acids would undergo decarboxylation to yield 2-pyrazinones. It is worth noting that these pyrazinones were only identified in the diglycine, triglycine and tetraglycine, but not in the free glycine system.

Table II shows the pyrazinones identified in the reaction of glucose with either gly-leu or leu-gly. It was found that the dipeptides, gly-leu and leu-gly, produced the same type of pyrazinones. Quantitatively, both dipeptides produced similar amounts of pyrazinones. The only possible explanation for this phenomenon is that the gly-leu dipeptide is in equilibrium with the leu-gly dipeptide.

Comparison of the Maillard Reaction of Glucose with Glutathione and a Mixture of their Constituent Amino Acids (glu, cys and gly)

It is known from our previous study (20) that the release of hydrogen sulfide was much faster than ammonia during the thermal degradation of glutathione (γ-Glu-Cys-Gly) in an aqueous solution. However, the release of both hydrogen sulfide and ammonia from cysteine was fast and produced four times as many volatiles as glutathione under the same conditions (20). On the other hand, when cysteine and glutathione reacted with 2,4-decadienal, respectively, the yield of volatile generation became almost identical. This phenomena was attributed to the fact that carbonyls, such as 2,4-decadienal and their retroaldolization products, catalyzed the ammonia released from glutathione via the formation of the Schiff-base (21).

We compared the Maillard reaction of equimolar solutions of glucose with glutathione (G-G) and its constituent amino acids (G-GCG) in an aqueous medium. Each reaction solution was adjusted to pH 7.5 and heated for one hour at 180°C. The reaction mass was simultaneously solvent-extracted and steam-distilled by using diethyl ether with a Likens-Nickerson apparatus. The distillates were dried over anhydrous sodium sulfate and concentrated with a Kuderna-Danish apparatus to a final volume of 0.5 mL. The concentrated samples were analyzed by GC/MS as described previously (22). A more roasted and nutty aroma was observed in the G-GCG system than in the

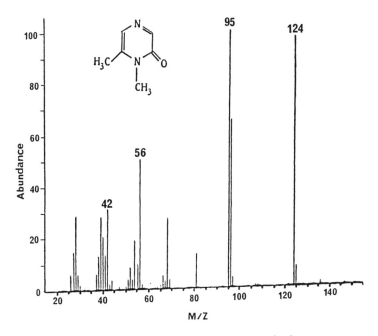

Figure 1. Mass spectrum of 1,6-dimethyl-2(1H)-pyrazinone.

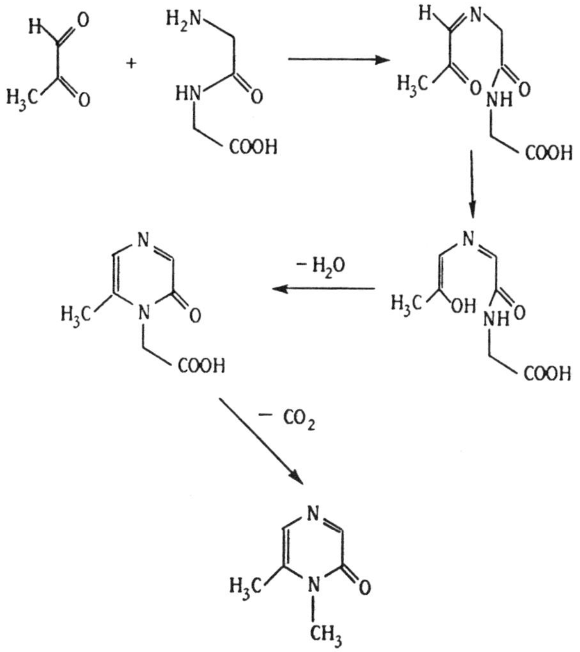

Figure 2. Mechanism for the formation of 1,6-dimethyl-2(1H)-pyrazinone from pyruvaldehyde and diglycine.

Table II.   The Amount of Pyrazinones Generated from the
Reaction of Glucose with Either
Gly-leu or Leu-gly

	MW	I_k	GL	LG
			mg/mole peptide	
1-methyl-3-isobutyl-2(1H)-pyrazinone	166	1397	351.41	433.59
1-isopentyl-2(1H)-pyrazinone	166	1432	185.58	129.95
1,6-dimethyl-3-isobutyl-2(1H)-pyrazinone	180	1562	1629.69	1234.09
1,5,6-trimethyl-3-isobutyl-2(1H)-pyrazinone	194	1638	84.17	83.54

G-G system. The identification and quantification of volatiles generated in these two systems are listed in Table III according to
their chemical classifications.
     It is interesting to note that about the same quantity of fu-
rans were produced by the G-G and G-GCG systems, however, the amino
acids mixture yielded 5.6 times the amount of carbonyl compounds
than the glutathione system. Furans are the cyclization products of
the sugar-derived Maillard intermediates and the carbonyl compounds
are generally derived from the fragmentation of sugar. The higher
contents of the free amino groups in the G-GCG system may favor the
sugar degradation reaction to yield carbonyl compounds such as 3-
hydroxy-2-pentanone and 2-hydroxy-3-pentanone.
     The generation of heterocyclic compounds such as pyrazines,
thiazoles and thiophenes from the G-GCG system had a much higher
yield than from the G-G system. The amino acids mixture produced
14 times more pyrazines than the glutathione when reacted with
glucose.
     The trithiolanes and tetrathianes identified are well-known
interaction products of hydrogen sulfide and acetaldehyde, both of
which are the decomposition products of cysteine. The G-GCG system
produced 32 times more of these cyclic polysulfides. Although glutathione is efficient in releasing hydrogen sulfide, it may not be
a good precursor for the generation of acetaldehyde.

Table III. Volatile Compounds Identified from the Interaction
of Glucose with Glutathione or Glutathione's
Constituent Amino ACids (Glu, Cys, Gly) in an Aqueous
Solution at pH 7.5 and 180°C

Compounds	MW	I^*_k	mg/mol. Glutathione	mg/mol. Glu+Cys+Gly
Carbonyls				
diacetyl	86	561	1.83	4.64
2-butanone	72	572	4.34	72.53
ethyl acetate	88	599	15.56	-
hydroxyacetone	74	-	-	t
2-pentanone	86	667	-	t
2.3-pentanedione	100	668	3.60	t
acetoin	88	680	0.82	-
3-penten-2-one	84	719	0.31	t
1-mercapto-2-propanone	90	741	-	1.28
2,4-pentanedione	100	758	-	t
3-hydroxy-2-pentanone	102	773	-	41.99
2-hydroxy-3-pentanone	102	782	-	18.28
3-mercapto-2-butanone	104	787	-	t
2,4-hexanedione	114	859	-	2.22
4-hydroxy-3-hexanone	118	865	-	1.30
2-cyclohexenone	96	904	-	5.79
Total Amount			26.46	148.03
Furans				
2-methylfuran	82	-	-	t
2,5-dihydrofuran	70	-	-	t
2,5-dimethylfuran	96	699	-	0.54
2,3-dihydro-4-methylfuran	84	-	-	t
2-methyltetrahydro-furan-3-one	100	776	9.26	17.08
furfural	96	809	9.95	1.19
furfuryl alcohol	98	835	0.88	6.52
2-acetylfuran	110	882	23.12	5.88
2,5-dimethyl-3(2H)-furanone	112	922	-	1.15
5-methylfurfural	110	934	111.01	111.46
5-methyl-2-acetylfuran	124	1019	2.13	-
1-(2-furyl)-1,2-propanedione	138	1034	2.61	t
Total Amount			158.96	143.82
Thiophenes				
thiophene	84	648	-	19.38
2,3-dihydrothiophene	86	-	-	t
2-methylthiophene	98	756	21.71	64.96
3-methylthiophene	98	760	-	17.08
2-ethylthiophene	112	844	1.69	0.35
2,5-dimethylthiophene	112	853	0.61	t
2,3-dimethylthiophene	112	877	0.81	t
tetrahydrothiophen-3-one	102	912	19.43	90.73
5-methyltetrahydro-thiophene-3-one	116	946	1.52	5.50

Continued on next page

Table III. Continued

| Compounds | MW | I^*_k | mg/mol. | |
			Glutathione	Glu+Cys+Gly
Thiophenes Continued				
2-methyltetrahydrothiophene-3-one	116	953	9.74	55.64
thiophene-2-carboxyaldehyde	112	958	4.40	38.42
thiophene-3-carboxyaldehyde	112	967	9.95	12.05
1-(2-thienyl)-ethanethiol	130	1035	-	6.13
2-acetylthiophene	125	1049	11.20	150.09
methylformylthiophene	126	1054	2.16	30.56
3-acetylthiophene	126	1058	24.31	177.93
5-methyl-2-formylthiophene	126	1086	11.49	27.54
3-methyl-2-formylthiophene	126	1089	5.20	65.73
5-methyl-2-acetylthiophene	140	1128	8.85	31.37
3-methyl-2-acetylthiophene	140	1135	3.34	3.51
2-(1-propionyl)-thiophene	140	1153	1.31	-
methylacetylthiophene	140	1174	-	34.70
thenio-[3,2-B]-thiophene	140	1185	-	60.67
3-methyl-2-(oxopropyl)-thiophene	154	1192	2.59	-
2,5-dimethyl-4-hydroxy-3(2H)-thiophene	144	1195	-	2.98
thieno-[2,3-C]-pyridine	135	1213	-	2.75
5-methylthieno-[2,3-B]-thiophene	154	1282	1.45	20.11
5-ethylthieno-[2,3-D]-thiophene	168	1380	-	6.58
Total Amount			141.76	924.76
Thiazoles				
thiazole	85	707	9.53	95.97
2-methylthiazole	99	783	1.29	t
4-methylthiazole	99	795	2.57	2.00
5-methylisothiazole	99	820	1.09	t
5-methylthiazole	99	827	-	3.09
2,5-dimethyl-thiazole	113	860	-	4.44
5-ethylthiazole	113	-	-	t
3,5-dimethylisothiazole	113	-	-	t
trimethylthiazole	127	977	0.45	9.35
2-acetylthiazole	127	988	21.68	39.56
trimethylisothiazole	127	-	-	t

Table III. Continued

Compounds	MW	I*$_k$	mg/mol. Glutathione	mg/mol. Glu+Cys+Gly
Thiazoles Continued				
4-methyl-2-acetyl-thiazole	141	1083	0.88	6.94
Total Amount			37.49	161.35
Pyrazine				
pyrazine	80	710	-	113.46
methylpyrazine	94	799	13.09	73.44
2,5-dimethyl-pyrazine	108	888	8.19	107.41
2-ethylpyrazine	108	893	1.06	15.31
2,3-dimethypyrazine	108	897	2.52	16.65
2-methyl-5-ethyl-pyrazine	122	977	-	5.68
trimethylpyrazine	122	979	3.91	47.06
2-methyl-6-ethyl-pyrazine	122	984	-	7.82
3,6-dimethyl-2-ethylpyrazine	136	1059	-	10.01
5,6-dimethyl-2-ethylpyrazine	136	1065	t	1.68
Total Amount			28.77	398.52
Pyridines				
pyridine	79	-	-	t
2-methylpyridine	93	798	-	5.20
Total Amount				5.20
Other Sulfur-containing Compounds				
2-pentanethiol	104	838	-	1.69
4H-tetrahydrothio-pyran-4-one	116	1011	0.44	2.71
3,5-dimethyl-1,2,4-trithiolane	152	1103	1.49	89.93
3,5-dimethyl-1,2,4-trithiolane	152	1110	4.40	101.23
3,6-dimethyl-1,2,4,5-tetrathiane	184	1345	-	10.12
3,6-dimethyl-1,2,4,5-tetrathiane	184	1352	-	5.69
4,6-dimethyl-1,2,3,5-tetrathiane	184	1368	0.70	3.79
4,6-dimethyl-1,2,3,5-tetrathiane	184	1396	-	0.38
Total Amount			7.03	215.54

[a] t = trace

* I_k = linear retention indices were obtained by using paraffin standards on a nonpolar fused silica capillary column (60 m x 0.25 mm [i.d.]); 0.25 μm thickness; DB-1; J&W Scientific).

Acknowledgements

New Jersey Agricultural Experiment Station Publication No. D-10544-14-91 supported by State funds, and the Center for Advanced Food Technology. The Center for Advanced Food Technology is a member of the New Jersey Commission for Science and Technology. We thank Mrs. Joan B. Shumsky for her secretarial aid.

Literature Cited

1.   Hodge, J. E. *J. Agric. Food Chem.* **1953,** *1*, 928-43.
2.   Shigematsu, H.; Shibata, S.; Kurata, T.; Kato, H.; Fujimaki, M. *J. Agric. Food Chem.* **1975,** *23*, 233-37.
3.   Tressl, R.; Rewicki, D.; Helak, B.; Kamperschroer, H.; Martin, N. *J. Agric. Food Chem.* **1985,** *33*, 919-23.
4.   Tressl, R.; Rewicki, D.; Helak, B.; Kamperschroer, H. *J. Agric. Food Chem.* **1985,** *33*, 924-28.
5.   Tressl, R.; Helak, B.; Koppler, H.; Rewicki, D. *J. Agric. Food Chem.* **1985,** *33*, 1132-37.
6.   Tressl, R.; Grunewald, G.K.; Kersten, E.; Rewicki, D. *J. Agric. Food Chem.* **1985,** *33*, 1137-42.
7.   Baltes, W.; Bochmann, G. *J. Agric. Food Chem.* **1987,** *35*, 340-46.
8.   Shu, C-K; Ho, C-T *J. Agric. Food Chem.* **1988,** *36*, 801-03.
9.   Zhang, Y.; Ho, C-T. *J. Agric. Food Chem.* **1991,** *39*, 760-63.
10.   Hartman, G. J.; Scheide, J. D.; Ho, C-T *Perf. Flav.* **1984,** *8(6)*, 31-6.
11.   Olsson, K.; Pernelmalm, P. A.; Theander, O. *Acta Chem. Scand.* **1978,** *B32*, 249-56.
12.   Hayase, F.; Kim, S. B.; Kato, H. *Agric. Biol. Chem.* **1985,** *49*, 2337-41.
13.   Takahashi, K.; Tadenuma, M.; Kitamoto,K.; Sato, S. *Agric. Biol. Chem.* **1973,** *38*, 927-32.
14.   Mabrouk, A. F. In *Phenolic, Sulfur and Nitrogen Compounds in Food Flavors*; Charalambous, G.; Katz, I., Eds.; Acs Symp. Ser. 26; American Chemical Society: Washington, DC, **1976**; p. 146-83.
15.   Manley, C. H.; McCann, J. S.; Swaine, Jr., R. L. In *The Quality of Foods and Beverages*; Charalambous, G.; Inglett, G., Eds.; Academic Press: New York, **1981,** *Vol. 1*; p 61-82.
16.   Chuyen, N. V.; Kurata, T.; Fujimaki, M. *Agric. Biol. Chem.* **1972,** *37*, 327-34.
17.   Rizzi, G. P. In *Thermal Generation of Aromas*; Parliment, T. H.; McGorrin, R. J.; Ho, C-T, Eds.; ACS Symp. Ser. 409; American Chemical Society: Washington, DC, **1989**; p 285-301.
18.   Oh, Y-C; Shu, C-K; Ho, C-T *J. Agric. Food Chem.* **1991,** *39*, 1553-54.
19.   Hayase, F.; Kato, H.; Fujimaki, M. *Agric. Biol.Chem.* **1975,** *39*, 741-42.
20.   Zhang, Y.; Chien, M.; Ho, C-T. *J. Agric. Biol. Chem.* **1988,** *36*, 992-96.
21.   Zhang, Y.; Ho, C-T *J. Agric. Food Chem.* **1989,** *37*, 1016-20.
22.   Zhang, Y.; Ho, C-T *J. Agric. Food Chem.* **1991,** *39*, 760-63.

RECEIVED December 18, 1991

Chapter 16

Meat Flavor Generation from Cysteine and Sugars

K. B. de Roos

PFW (Nederland) BV, P.O. Box 3, 3800 AA Amersfoort, Netherlands

Flavor generation from cysteine−sugar systems is complicated by the low reactivity of these systems and the instability of the resulting meaty flavor. The reactivity of the cysteine−sugar systems was found to be related to the concentration of the neutral thiazolidinecarboxylic acid intermediates in solution. The higher the concentration of the neutral intermediates, the higher the reactivity. In the anionic form, the reactivity of the thiazolidinecarboxylic acids is very low.

In more complex amino acid−sugar systems, the presence of cysteine always resulted in inhibition of browning due to the formation of relatively stable thiazolidines. The inhibition of the first steps of the Maillard reaction could be circumvented by the use of Amadori compounds. Mixtures of Amadori compounds with cysteine produced more browning and stronger aromas than equivalent mixtures of the corresponding amino acids and sugars.

The role of cysteine in the development of meat flavor has long been recognized (*1−3*). Cysteine contributes to meat flavor by its participation in the Maillard reaction (Figure 1). The dicarbonyls formed during this reaction catalyze the Strecker degradation of cysteine to form mercaptoacetaldehyde, acetaldehyde and hydrogen sulfide as the primary degradation products (Figure 2). The formation of these Strecker degradation products is the start of a cascade of reactions that leads to the formation of a wide range of other meat flavor compounds (*4−6*).

Unfortunately, there are two factors that complicate meat flavor development from cysteine. One of these is the inhibition of the Maillard reaction by cysteine. Cysteine has been reported to inhibit non−enzymatic browning in miso (*7*), milk (*8*), egg (*9*) and fruit juices (*10*). The inhibitory effect of cysteine has been attributed to the formation of stable thiazolidines (*9*) and thio(hemi)acetals or thio(hemi)ketals (*10*) upon reaction with sugars and the (di)carbonyls that are formed during the Maillard reaction.

0097–6156/92/0490–0203$06.00/0
© 1992 American Chemical Society

CARBOHYDRATES

| + Amino Acids

AMADORI + HEYNS COMPOUNDS

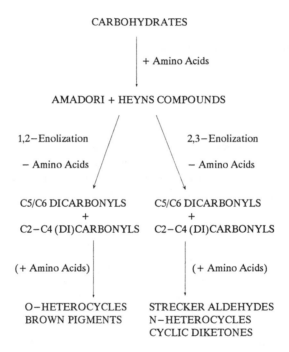

1,2−Enolization   2,3−Enolization

− Amino Acids   − Amino Acids

C5/C6 DICARBONYLS   C5/C6 DICARBONYLS
+   +
C2−C4 (DI)CARBONYLS   C2−C4 (DI)CARBONYLS

(+ Amino Acids)   (+ Amino Acids)

O−HETEROCYCLES   STRECKER ALDEHYDES
BROWN PIGMENTS   N−HETEROCYCLES
                                          CYCLIC DIKETONES

Figure 1.   Major Maillard reaction pathways to the formation of flavor
compounds and brown pigments.

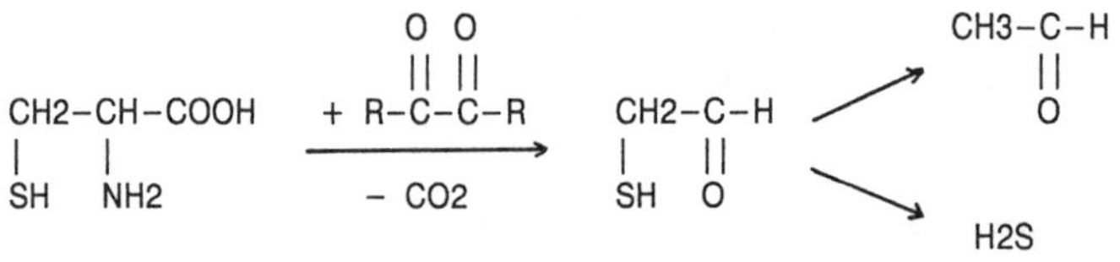

CH2−CH−COOH   + R−C−C−R   CH2−C−H
| | | ||
SH   NH2   − CO2   SH O

with the dicarbonyl reagent having structure with two O double-bonded carbons, yielding CH_3-C-H with double bond O, and H_2S.

Figure 2.   Primary products resulting from Strecker degradation of cysteine.

The second complicating factor is the low stability of meat flavor (*11*). Generation of unstable aromas requires the use of reactive precursor systems that develop the full flavor already under mild conditions, preferably *in situ* during the heating of the meat just before its consumption. In this study we will focus on the problem of how to increase the reactivity of cysteine in Maillard reaction systems.

Factors that Affect the Rate of the Maillard Reaction

The following factors have been found to affect the reactivity of the Maillard reaction:

Proportion and Nature of the Reactants. In general, the sugar has less influence on the sensory properties of the final flavor than the amino acid (*12*). Therefore, use of reactive sugars appears to be an attractive approach to enhancing the rate of meat flavor generation without compromising too much on flavor quality. This would imply that pentoses are to be preferred over hexoses.

Water Activity / Moisture Content. Maximum rate of Amadori compound formation is observed at water activities of 0.30−0.75 depending on the temperature of the system (*13*). Highest flavor yields are obtained at a water activity of 0.72 (*14*). Meat flavor generation from beef at different water activities has been shown to result in different flavor (*15*).

pH. In general, browning is favored by high pH. Often a maximum is found at about pH 10 (*16*). The pH affects flavor by influencing the yields of the various flavor compounds in a different way (*17*).

Presence of Buffer Salts. The Maillard reaction is subject to general acid catalysis. This means that the reaction rate is dependent on the concentration of all acid species in solution rather than on the hydrogen ion concentration. Phosphate and lactate have been found to be effective catalysts of the Maillard reaction (*18−20*).

Time / Temperature. Time and temperature of heating are known to strongly affect flavor development in Maillard reaction systems. Very different flavors can be produced from the same reaction system simply by variation of temperature and time of heating (*17*). Generation of unstable aromas like that of meat requires the use of moderate reaction temperatures and times.

Experimental

Analytical Methods

High Performance Liquid Chromatography. Quantitative analyses of Maillard reaction intermediates, amino acids and sugars were performed on a Spectra Physics HPLC System equipped with a refractive index detector. Analyses were run on a

Hypersil APS (aminopropylsilane) HPLC column using 65:35 (v/v) acetonitrile −
0.01 M sodium acetate pH 7 as the eluent.

Measurement of Browning. After cooling, the solutions were filtered using
Millex HV filters. If necessary, dark brown solutions were diluted with the
appropriate amount of water for accurate measurement of browning. Absorbance
was measured at 490 nm with a Varian Model CARY 1 UV−Visible
Spectrophotometer. The absorbance was adjusted for differences in the initial sugar
concentrations to provide a general measure for the browning generated per mole of
sugar (Absorbance / M sugar, where M = initial sugar concentration in moles/liter).

Reaction Conditions

Reaction of Cysteine with Sugar at Room Temperature. Sugar (25 mmol) and
cysteine (25 mmol) were dissolved in 20 ml of water. Then 5 ml of water or 5 ml of
5 M sodium hydroxide solution were added. The final volumes of the resulting
solutions varied from 29.3 to 29.6 ml which corresponds with a sugar concentration
of 0.85 M. The solids content varied from 21 to 26% (w/w). The solutions were
allowed to stand at room temperature in closed brown bottles. Samples were taken
after 3, 7.5, 26 and 49 hrs for subsequent HPLC analysis.

Heating of Sugars and Amino Acids at 90 °C. Sugar (25 mmol), cysteine
(0 or 25 mmol), other amino acid (0 or 25 mmol) and phosphate (0 or 50 mmol) were
dissolved in 20 ml of water. Then 5 ml of water or sodium hydroxide solution (0 or
5 M) were added. The resulting 0.85 M cysteine−sugar solutions were heated with
stirring in a round bottomed flask placed in a water bath maintained at 90 °C.
Samples were taken after 1, 2 and 4 hrs of heating. After cooling in ice water,
pH and absorbance were measured.

Concentrated solutions (30% water) were prepared from the dilute solutions by
evaporation of the water *in vacuo*.

Reactions in Propylene Glycol−Water at 120 °C. Sugar (0 or 6 mmol),
cysteine (7.5 mmol), other amino acid (0, 6 or 30 mmol), and Amadori compound
(0 or 6 mmol) were dissolved in 6 ml of water. Propylene glycol (12 ml) was added and
the volume was adjusted to 20 ml by addition of propylene glycol−water 2:1. After
mixing, the clear solution was distributed over four 5 ml reaction vials (4 x 4 ml
aliquots). The vials were heated in an oven thermostated at 120 °C for 0.25, 0.5,
1.0 and 1.5 hrs, respectively.

Preparation of Amadori Compounds. Ketose−amino acids and
di−D−fructose−glycine were prepared and isolated as described by Anet (*21, 22*).

Organoleptic Evaluation. The flavor character and intensity of the reaction
products were evaluated in a reconstituted bouillon base by an experienced flavorist.

Discussion of Results

Measurement of Rate of the Maillard Reaction.
Initial rate of browning was used as a general measure for the reactivity of the Maillard reaction systems. In general, a good relationship was found to exist between initial rate of browning and flavor strength provided that variations in pH and flavor precursor composition are limited. Best meat flavors were obtained at pH < 7. In some cases, concentrations of the Maillard reaction intermediates were also determined using HPLC.

Reactivity of Cysteine–Sugar Systems

Reaction at Room Temperature.
Figure 3 shows the rate of thiazolidine formation in 0.85 M equimolar mixtures of cysteine with D–fructose (FRU), D–glucose (GLU), D–xylose (XYL), and D–ribose (RIB) at low and high pH. The rate of formation was measured by HPLC. The results demonstrate that at low pH equilibrium between cysteine, sugar and thiazolidinecarboxylic acid has been established after about 25 hrs. At high pH, the equilibrium establishes at a lower rate but at the end the conversion is higher. The completeness of the conversion is a function of the type of the sugar. It is highest with pentoses.

Analogy with the cysteine–formaldehyde system (*23–24*) suggests that the mechanism of the reaction of cysteine with sugar to thiazolidine is as shown in the scheme of Figure 4 (the predomunant ionic forms at neutral pH are shown). According to this mechanism, the first step, the general acid–catalyzed condensation of the amine with the carbonyl group is rate determining at low pH, whereas the dehydration of the carbinolamine is rate determining at alkaline pH.

Since the thiazolidinecarboxylic acids are stronger acids than cysteine, pH drops with time, thus counteracting completion of the reaction. For the same reason, the thiazolidines are among the first compounds that are converted to anionic form in alkaline medium. As a consequence, the equilibrium shifts in alkaline medium from the side of the starting materials to the side of the thiazolidinecarboxylate anion **6**.

Browning at Elevated Temperature.
Figure 5 compares the rates of browning in dilute and concentrated equimolar cysteine–sugar systems at 90° C as function of pH. The results clearly demonstrate that in concentrated medium rates of browning per mole of sugar are higher than in dilute medium. At low and neutral pH (0 and 0.5 equivalents of NaOH added), reaction rates differ by a factor of more than 10.

At low and neutral pH, ribose is the most reactive sugar followed by xylose, glucose and fructose. So, the sugars show here the same order of reactivity as they do in the thiazolidine formation. This suggests that high conversion to thiazolidinecarboxylic acid is a prerequisite for rapid browning and flavor development. It also suggests that the neutral thiazolidinecarboxylic acid 5 (or the cationic Schiff base 4 which is in fast equilibrium with 5) is the most likely intermediate to the formation of brown pigments and flavor compounds.

In *dilute* cysteine–sugar systems, increase of the pH to values higher than 9 results in a complete reversal of the reactivity. This is due to a substantial decrease of

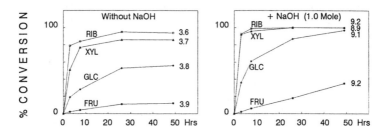

Figure 3. Rate of thiazolidine formation in equimolar 0.85 M cysteine–sugar
solutions at room temperature as a function of the amount of added
sodium hydroxide (in moles per mole of cysteine). Numbers on the
graphs refer to the pH at the end of the reaction period.

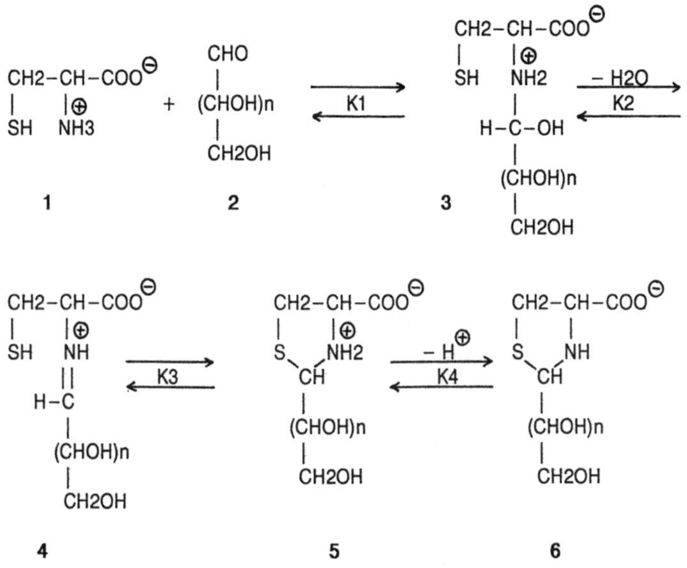

Figure 4. Major equilibria in cysteine–sugar systems at neutral pH.

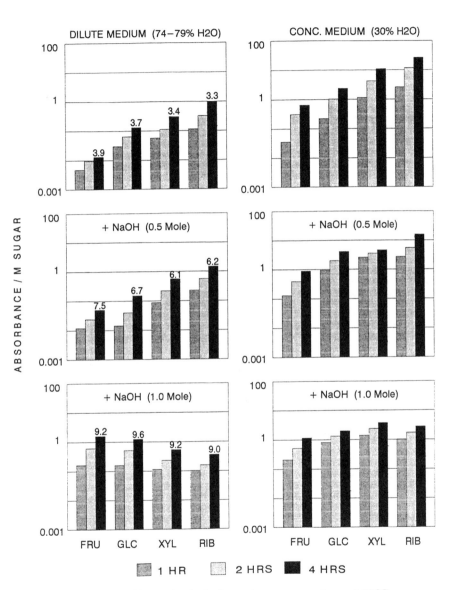

Figure 5. Rates of browning in 1 : 1 cysteine−sugar systems at 90 °C as a function of water content and the amount of added sodium hydroxide (in moles per mole of cysteine). Numbers above the bars refer to the pH after 4 hrs of heating.

the concentrations of the reactive intermediates **4** and **5** in the systems with highest thiazolidine concentrations. In these systems, the addition of an equimolar amount of sodium hydroxide results in an almost complete conversion of **4** and **5** into the much more stable thiazolidinecarboxylate anion **6**. As a consequence, the rate of browning decreases. However, if the equilibrium is mainly on the side of starting materials, a shift of equilibrium K4 will not result in a proportionate decrease of the intermediates **4** and **5** , because part of the initial decrease will be undone by a shift of the equilibria from starting materials **1** and **2** to neutral thiazolidinecarboxylic acid **5**. This is the reason that in cysteine−hexose systems rates of browning continue to increase with increase of pH, whereas in cysteine−pentose systems under these conditions inhibition of browning is most predominant.

In *concentrated* medium, inhibition of browning becomes already predominant at low and intermediate pH values. Both ribose and xylose show now highest rate of browning in weakly acidic medium, while maximum rate of browning in cysteine−glucose systems occurs at about neutral pH. Only fructose still shows highest rate of browning in alkaline medium.

The increase of inhibition of browning at increase of the reactant concentrations can be explained in a way analogous to the explanation of inhibition at high pH. In concentrated medium, the conversion of cysteine and sugar to thiazolidine is more complete than in dilute medium. And as we have seen before, the more complete the conversion to thiazolidine, the higher the decrease of the reactive intermediates **4** and **5** at increase of pH and the higher the inhibition of browning.

Catalysis by Phosphate. Figure 6 shows the effect of phosphate buffer on browning in cysteine−sugar systems. Interpretation of the differences in browning between buffered and unbuffered systems is slightly complicated by the differences in change of pH during heating. On an average, the presence of phosphate enhances the rate of browning by a factor of 10 over the whole pH range. A corresponding enhancement of meat flavor development is observed as well.

"Complex" Amino Acid−Sugar Systems

Pathways to Browning and Flavor Development. Addition of other amino acids to cysteine−sugar systems may be expected to result in enhanced browning and flavor development for two reasons. First of all, because the added amino acid will catalyze the cysteine−sugar reaction ("thiazolidine pathway"). Secondly, because a Maillard reaction will take place between the sugar and the added amino acid ("Amadori pathway").

In general, the rate of browning in cysteine−sugar systems is much lower than in other amino acid−sugar systems (*25*). For example, the rates of browning in the glycine−fructose systems (Figure 7) are up to 2000 times higher than those in the cysteine−fructose systems (Figure 5). The low reactivity of cysteine−sugar systems has to be attributed to the high stability of the thiazolidines as compared to the Amadori compounds in glycine−sugar systems. Another cause of the low reactivity of cysteine−sugar systems is the interaction of cysteine with the carbonyl compounds that are formed in the course of the Maillard reaction.

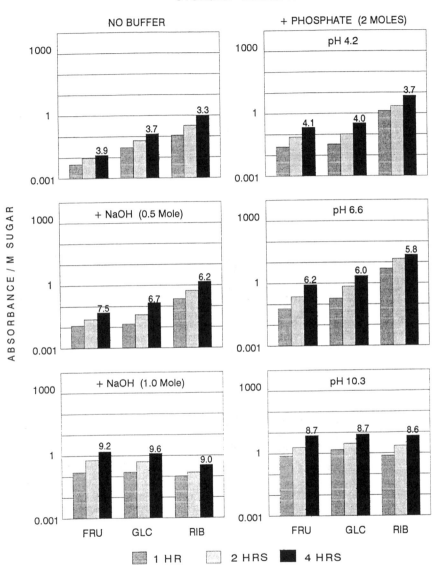

Figure 6. Effect of phosphate on rate of browning in cysteine−sugar systems at 90 °C as function of pH. The pH of the phosphate buffer is indicated in the upper part of the graph. Numbers above the bars refer to the pH after 4 hrs of heating.

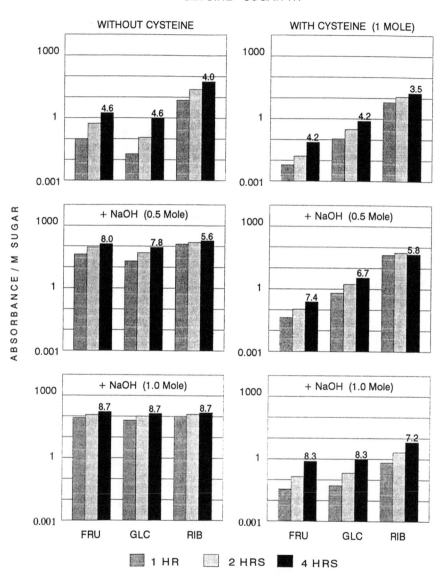

Figure 7.   Rates of browning in 0.85 M glycine−sugar and cysteine−glycine−sugar systems at 90 °C as a function of the amount of added sodium hydroxide (in moles per mole of glycine). Numbers above the bars refer to the pH after 4 hrs of heating.

In spite of the large differences in the reactivity between cysteine and the other amino acids, it is difficult to say beforehand which pathway will make the highest contribution to browning and flavor development in systems containing other amino acids in addition to cysteine. The reason is that the cysteine−sugar reaction is catalyzed by the amino acids, whereas the amino acid−sugar reaction is inhibited by cysteine. To get a better insight into the preferred reaction pathways to meat flavor development in amino acid−sugar systems, a few additional experiments were carried out.

"Thiazolidine Pathway". Figure 7 (right part) shows the browning in cysteine−sugar systems in the presence of an equimolar amount of glycine. Comparison with the cysteine−sugar systems without glycine (Figure 5, left part) demonstrates that the effect of glycine on browning is highest at neutral pH. At low and high pH, the effect of glycine on the initial rate of browning is sometimes almost negligible (in particular, in the cysteine−fructose systems). This suggests that, in general, the catalytic effect of glycine on the cysteine−sugar reaction is low. The relatively large increase of the rate of browning in the cysteine−ribose systems has most likely to be attributed to the Maillard reaction between glycine and ribose ("Amadori pathway").

"Amadori Pathway". Comparison of the glycine−sugar systems with the glycine−cysteine−sugar systems (Figure 7) shows that inhibition of browning by cysteine is highest with fructose. At high pH, rates of browning differ by a factor of more than 2000. At first sight, this result is unexpected, since fructose does not form appreciable amounts of thiazolidine. A possible explanation is that inhibition of browning occurs at a later stage of the Maillard reaction and that the degradation products of the hexoses are more susceptible to inhibition by cysteine than those derived from pentoses. Part of the differences in browning can also be contributed to the higher browning potential of the cysteine−pentose sytems.

HPLC analysis of the glycine−cysteine−sugar systems demonstrated that Amadori compounds of glycine are being formed during heating. This supports the previous assumption that the "Amadori pathway" is likely to contribute to browning in these systems. Apart from being flavor precursors by themselves (*26, 27*), Amadori compounds can serve as precursors of reactive dicarbonyl compounds that catalyze the Strecker degradation of cysteine. If indeed the Strecker degradation of cysteine is a key step in its conversion to meat flavor (**4−6**), it may be expected that the rate of meat flavor development will be enhanced by use of Amadori compounds. To check this hypothesis, a few experiments with Amadori compounds carried out under simulated roasting conditions (Table I).

In system A, substitution of proline and ribose by an equivalent amount of the corresponding Amadori compound was found to enhance browning by a factor of 7 − 13 depending on the period of heating. Meat flavor development is enhanced as well. In fact, no meat flavor development is observed at all without the use of Amadori compounds. In the presence of an excess of proline (system B), the differences in browning and meat flavor development are much less pronounced.

Table 1. Effect of Amadori compounds on browning and flavor development in
2:1 propylene glycol−water model systems at 120 °C

	Composition (mol/l) *				Time	A490	Flavor
	CYS	RIB	PRO	RIB−PRO	(Hrs)		
A1	0.375	0.3	0.3	−	1.0	1.4	weak, roast
					1.5	3.4	roast
A2	0.375	−	−	0.3	0.5	15.8	roast, meaty, cereal
					1.0	18.4	meaty, roast
					1.5	23.5	meaty, slightly roast
B1	0.375	0.3	1.5	−	0.5	10.4	roast, cereal
					1.0	12.7	meatier but roast
					1.5	9.4	meaty, slightly cereal
B2	0.375	−	1.2	0.3	0.5	13.8	similar
					1.0	25.7	to A2 but
					1.5	34.1	stronger
	CYS	GLC	GLY	FRU−GLY			
C1	0.375	0.3	1.5	−	0.5	1.1	weak, sulfury
					1.0	2.3	sulfury, onion, roast
					2.0	2.5	roast, onion
C2	0.375	−	1.2	0.3	0.5	1.3	weak
					1.0	2.5	meaty, hint of onion
					2.0	4.5	heavy meaty, slightly onion
	CYS	GLC	GLY	Di−FRU −GLY			
D1	0.375	−	1.35	0.15	0.25	18.7	weak, sulfury
					0.5	9.2	sulfury, meaty, H2S
					1.0	7.6	sulfury, meaty, H2S
					2.0	6.6	meaty, hint of onion

* CYS = L−cysteine; PRO = L−proline; GLY = glycine; RIB = D−ribose
GLC = D−glucose; RIB−PRO = D−ribulose−L−proline
FRU−GLY = D−fructose−glycine; DiFRU−GLY = di−D−fructose−glycine

For the glycine−cysteine−glucose system C (excess of glycine), the effect of using an Amadori compound is also less convincing, in particular, at prolonged reaction times. However, with difructose−glycine (system D) rate of browning is extremely fast. Maximum browning is already achieved within 15 minutes. Unfortunately, a comparable increase in the rate of flavor formation is not observed. Since difructose−glycine is known to decompose rapidly into fructose−glycine and 3−deoxyglucosulose (28), it may be concluded that 3−deoxyglucosulose is a precursor of brown pigments rather than of meaty flavor. This does not mean that the osulose does not contribute at all to meat flavor development. In fact, Wedzicha and Edwards (29) found that reaction of cysteine with 3−deoxyhexosulose results already at room temperature in development of meaty flavor.

The above experiments clearly demonstrate that in amino acid−cysteine−sugar systems the "Amadori pathway" is likely to make a significant contribution to meat flavor development.

Conclusions

In simple cysteine−sugar systems, a positive relationship was found to exist between browning and the concentration of neutral thiazolidinecarboxylic acid in solution. Conversion of the neutral thiazolidine to the corresponding anion results in inhibition of browning and flavor development due to the much higher stability of the latter. In general, use of reactive sugars and buffer salts was found to be most effective in enhancing browning and meat flavor development in cysteine−sugar systems. With cysteine−pentose systems, the reactivity is also enhanced by a high solids content in combination with low pH.

In more complex amino acid−sugar systems, the presence of cysteine always results in inhibition of browning because the formation of the relatively stable thiazolidines prevents the reaction of sugars and sugar degradation products with other amino acids to brown pigments. In these systems two routes to meat flavor development exist. One route proceeds via neutral thiazolidine intermediates, the other via Amadori compounds.

The inhibition of the first steps of the Maillard reaction by cysteine could be circumvented by using Amadori compounds instead of amino acid−sugar mixtures. It was found that mixtures of Amadori compounds and cysteine produce stronger aromas than equivalent mixtures of the corresponding amino acids and sugars.

Acknowledgments

I thank C. Wolswinkel for technical assistance, A. van Delft for the organoleptic evaluations and A. Sarelse for assisting in the preparation of the manuscript.

Literature Cited

1. Macy, R.L.; Naumann, H.D.; Baily, M.E.   *J. Food Sci.* **1964,** *29,* 142
2. May, C.G.   *Food Trade Rev.* **1974,** *44,* 7

3. Morton, I.D.; Akroyd, P.; May, C.G. *Brit. Patent* 836694, **1960**
4. Kobayashi, N.; Fujimaki, M. *Agric. Biol. Chem.* **1965**, *29*, 698
5. Palamand, S.R.; Hardwick, W.A.; Davis, D.P. *Proc. Am. Soc. Brew. Chem.* **1970** , 56
6. Tressl, R.; Helak, B.; Martin, N. Kersten, E. In *Thermal Generation of Aromas*; Parliment, T.H.; McGorrin, R.J.; Ho, C.T.; Eds.; American Chemical Society; Washington, DC, 1989; p 156
7. Fujii, M.; Gondo, H.; Miyazaki, Y.; To, S. *Saga Daigaku Noguka Iho* **1973**, *79*, 6 ; *Chem. Abstr.* **1973** , *34*, 114133g
8. Arnold, R.G. *J. Dairy Sci.* **1969**, *52*, 1857
9. Lightbody, H.D.; Fevold, H.L, *Adv. Food Res.* **1948**, *1*, 149
10. Molnar−Perl, I.; Friedman, M. *J. Agric. Food Chem.* **1990**, *38*, 1648
11. Vercellotti, J.R.; Kuan, J.W.; Spanier, A.M.; St Angelo, A.J. in *Thermal Generation of Aromas* ; Parliment, T.H.; McGorrin, R.J.; Ho, C.T.; Eds.; American Chemical Society; Washington, DC, 1989; p 452
12. Shibamoto, T. In *Instrumental Analysis of Foods* ; Charalambous, G.; Inglett, G.; Eds.; Academic Press, Inc.; New York, 1983; Vol. 1, p 229
13. Eichner, K.; Laible, R.; Wolf, W. In *Properties of Water in Food,* Simatos, D.; Multon, J.L.; Eds.; Martinus Nijhoff Publishers, Dordrecht/Boston/Lancaster, 1985, p 191
14. Hartman, G.J.; Scheide, J.D.; Ho, C.T. *J. Food Sci.* **1984**, *49*, 607
15. MacLeod, G; Ames, J.M. *J. Food Sci.* **1987**, *52*, 42
16. Ashoor, S.H.; Zent, J.B. *J. Food Sci.* **1984**, *49*, 1206
17. Leahy, M.M.; Reineccius, G.A. In *Thermal Generation of Aromas;* Parliment, T.H.; McGorrin, R.J.; Ho, C.T.; Eds.; American Chemical Society; Washington, DC, 1989; p 196
18. Reynolds, T.M.; In *Symposium on Foods "Carbohydrates and Their Roles"* The Avi Publishing Co., Westpoint, Conn. 1969, p 219
19. Potman, R.P.; Van Wijk; T.A. in *Thermal Generation of Aromas* ; Parliment, T.H.; McGorrin, R.J.; Ho, C.T.; Eds.; American Chemical Society; Washington, DC, 1989; p 182
20. Saunders, J.; Jervis, F. *J. Sci. Fd. Agric.* **1966**, *17*, 245
21. Anet, E.F.L.J. *Austr. J. Chem.* **1957**, *10*, 193
22. Anet, E.F.L.J. *Austr. J. Chem.* **1959**, *12*, 280
23. Kallen, R.G. *J. Am. Chem. Soc.* **1971**, *93*, 6227
24. Kallen, R.G. *J. Am. Chem. Soc.* **1971**, *93*, 6236
25. Ames, J.M. *Chem. & Ind.* **1986**, 362
26. Mills, F.D.; Hodge, J.E. *Carbohydrate Res.* **1976**, *51*, 9
27. Van den Ouweland, G.A.M.; Peer, H.G.; Tjan, S.B. In *Flavor of Foods and Beverages* ; Charalambous, G.; Inglett, G.E.; Eds.; Academic Press, New York, 1978, p131
28. Anet, E.F.L.J. *Austr. J. Chem.* **1960**, *13*, 396
29. Wedzicha, B.L.; Edwards, A.S. *Food Chem.* **1991**, *40*, 71

RECEIVED December 18, 1991

Chapter 17

Analysis, Structure, and Reactivity of 3-Deoxyglucosone

H. Weenen and S. B. Tjan

Quest International, P.O. Box 2, NL 1400 CA Bussum, Netherlands

To achieve better control and understanding of selectivity in flavour formation by the Maillard reaction, the intermediate 3-deoxyglucosone was investigated. A method to purify and analyze glucosones by HPLC was developed, using silica gel linked α-cyclodextrin as the stationary phase. Quantities of more than 100 mg could be purified this way. The structure of 3-deoxyglucosone was studied using various ^{1}H-NMR and ^{13}C-NMR techniques. The reactivity of 3-deoxyglucosone was investigated, including its behaviour in aqueous acid and base. Pyrazine formation as a model reaction for sugar fragmentation was studied, using various hexoses, e.g. fructose, glucose, 1-^{13}C-glucose, and 3-deoxyglucosone.

One of the problems in using the Maillard reaction to generate flavour substances is its poor selectivity. Better control of selectivity could lead to both a larger number of olfactory targets and higher yields. Traditionally the result of a Maillard process is largely determined by the choice of the starting materials. However, the scope of the Maillard reaction would gain enormously if a finer control of all key steps generating a Maillard Reaction product were possible.

Of all Maillard reaction intermediates, glycosones appear to play the most important role in determining the nature of the end product. When a reducing sugar reacts with an amino acid, the first steps resulting in the formation of the glycosones seem relatively straightforward leading mainly to two glycosones. Glycosones, however, can react further to a very broad spectrum of products, mainly by condensation, fragmentation and rearrangement.

For glucose and fructose, glycosones of most importance are 1- and 3-deoxyglucosone. 1-Deoxyglucosone has been synthesized by Ishizu et al. (*1*)

0097–6156/92/0490–0217$06.00/0
© 1992 American Chemical Society

from 1-deoxyfructose, but it appears to be rather unstable, and little has been reported on its properties. 3-Deoxyglucosone (scheme 1, **1**) has been studied in more depth, especially by Anet (*2*), who synthesized it from difructoseglycine, and studied its conversion to 4-(hydroxymethyl)furfural (HMF) and metasaccharinic acid (MSA) (*3*). Kato et al. isolated several 3-deoxyglycosones as their bis-2,4-dinitrophenylhydrazones under Maillard reaction conditions (*4*), and established their role in melanoid formation and protein cross-linking (*5*). Khadem et al. (*6*) devised a convenient synthesis of 3-deoxyglycosones via their bis(benzoyl)hydrazones, which was later optimized by Madson and Feather (*7*). 3-Deoxyglucosone was also found to be a precursor of 5-methylpyrrole-2-carboxaldehyde, 6-methyl-3-pyridinol and 2-methyl-3-pyridinol, upon refluxing this osone with glycine in aqueous solution (pH3) (*8*).

The formation of glucosones from Amadori rearrangement products was investigated by Beck et al. using diaminobenzene (*9*), which can trap compounds containing an α-dicarbonyl moiety. Amadori rearrangement products derived from glucose and maltose were found to give mainly 1- and 3-deoxyglucosones, but also 4-deoxyglucosones. The role of the deoxyglycosones in the Maillard reaction in foods and in the human body has recently been excellently reviewed by Ledl (*10*).

Earlier it had been concluded that the formation of 1- versus 3-deoxyglucosone in the Maillard reaction is to some extent influenced by the pH and the basicity of the reacting amino acid (*11*). Under acidic conditions the formation of 3-deoxyglycosone is favoured, while under neutral or slightly alkaline conditions more 1-deoxyglycosone is formed. In a system where the pH is not regulated, the product distribution seems to indicate that 1- and 3-deoxyglucosone are formed at about equal rates (*12*). The nature of N- and S-heterocyclic Maillard products is also to a large extent determined by the precursor deoxyglycosones (*13*).

The question which in our opinion has so far not been answered satisfactorily, and which we are trying to address here, is how glycosones are cleaved to smaller fragments. The fragments are of particular importance because they react further to generate flavour substances such as pyrazines, thiazoles, carbocyclic compounds, etc. To answer that question 3-deoxyglucosone (3-deoxy-D-erythro-hexosulose) was selected as a model compound for glycosones in general. Its structure was studied by NMR, and its reactivity investigated, in particular the pyrazine formation reaction as a model for the fragmentation processes.

Experimental

HPLC. 3-Deoxyglucosone was analysed and purified using a Cyclobond-III column (Advanced Separation Technology, N.J., U.S.A.) and acetonitrile/water, 88/12, v/v, as the eluent. The column was regularly washed with pure water to remove cyclodextrin coming off the column. HMF was analysed on a BioRad fruit quality column using 0.002 N H_2SO_4 aq. as the eluent. Metasaccharinic acid was determined by anion chromatography in combination with a pulsed electrochemical detector (Dionex series 4500), eluent: 0.1 N NaOH aq. Samples in which metasaccharinic acid was measured, were also analysed for 3-deoxyglucosone, because 3-deoxyglucosone is converted to metasaccharinic acid at the pH of the eluent. A mixture of α- and β-metaglucosaccharinic acids was synthesized from 3-methoxyglucose (*14*), to be used as a standard.

NMR. NMR spectroscopy experiments included ^{1}H-NMR, ^{13}C-NMR, DEPT, COSY and J-resolved ^{1}H-NMR, at 100, 200 and 360 MHz. All spectra were taken in CDCl$_3$ or D$_2$O. ^{13}C-NMR spectra were taken at 30° with an accumulation time of 16 hours. The ^{13}C/^{12}C ratios were determined in duplicate, using integrated peak areas of the ^{13}C satellites and the uncoupled H-resonance peaks. ^{13}CH$_3$ incorporation in dimethylpyrazine was found to be about 26% (20-31%), and ^{13}CH incorporation in all pyrazines combined about 20% (14-26%).

Pyrazines. Pyrazines were obtained by heating a 1,2-propanediol solution of a hexose, asparagine and K$_2$HPO$_4$ at 130° for 1 hr. They were isolated by Likens-Nickerson extraction (*15*), and analysed by GC, using methoxypyrazine as the internal standard. Since the main dimethylpyrazine isomers observed (2,5- and 2,6-) were not, or poorly resolved by GC, we use the term dimethylpyrazine throughout the text, to indicate a mixture of these two isomers.

Results and Discussion

Analysis and structure of 3-deoxyglucosone

HPLC, UV and IR of 3-deoxyglucosone. No method to analyse glycosones by HPLC has been described. The only reports on chromatography of glycosones concern paper chromatography (*2*). Most HPLC packing materials which are commonly used to separate and analyse sugars proved useless in separating 3-deoxyglucosone, due to the reactivity and polarity of this compound. However silica gel linked α-cyclodextrin as the stationary phase in combination with acetonitrile/water as the eluent, allowed both preparative separation and analysis of 3-deoxyglucosone. Reasonably sensitive detection could be achieved by either refractive index detection (detection limit: 0.15 mg) or UV absorbance at 200 nm (detection limit: 0.01 mg). Chromatograms obtained by either RI detection or UV detection (200nm) showed no differences in shape and retention time for the peak representing 3-deoxyglucosone (Fig. 1). The chromatogram obtained by UV detection at 220 nm however, clearly shows a double peak for 3-deoxyglucosone. The UV spectrum of the first peak in the chromatogram (detection by UV absorption at 220 nm) was taken with a scanning UV detector, and showed only a very small maximum at 302 nm, in addition to a much stronger absorption between 200-225 nm (Figure 2). The IR spectrum of 3-deoxyglucosone (KBr) showed only a very small absorption in the 1700-1800 cm^{-1} region. Apparently 3-deoxyglucosone is mostly in the hemiacetal and/or hydrated form, but exists to a small extent in the monocarbonyl form.

NMR of 3-deoxyglucosone. ^{1}H-NMR and ^{13}C-NMR of purified 3-deoxyglucosone indicated that 3-deoxyglucosone, when dissolved in water, is present in about 5 major, and at least 6 minor isomers. No carbonyl absorptions could be detected by ^{13}C-NMR, indicating that 3-deoxyglucosone is almost exclusively in the (hemi)-acetal, (hemi)-ketal and/or hydrated form. The structures of all hypothetical cyclic isomers of 3-deoxyglucosone are indicated in scheme 1a and 1b. Scheme 1a lists the formal structure **1**, and the hydrates **2-10**. Hydrates are unlikely since they are generally only stable when

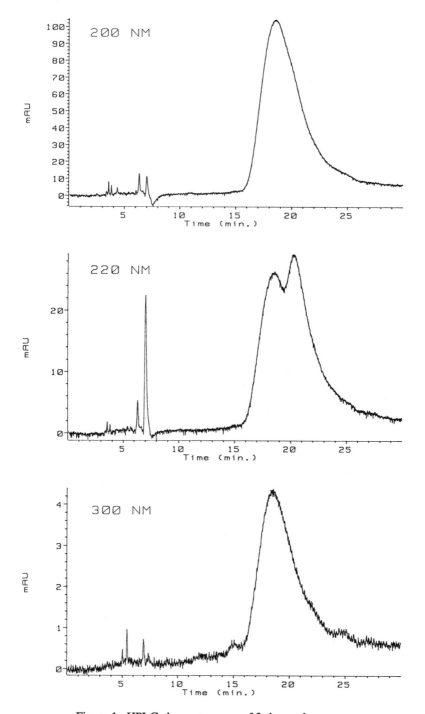

Figure 1. HPLC chromatograms of 3-deoxyglucosone

Scheme 1a. Formal and hypothetical hydrated structures of 3-deoxyglucosone.

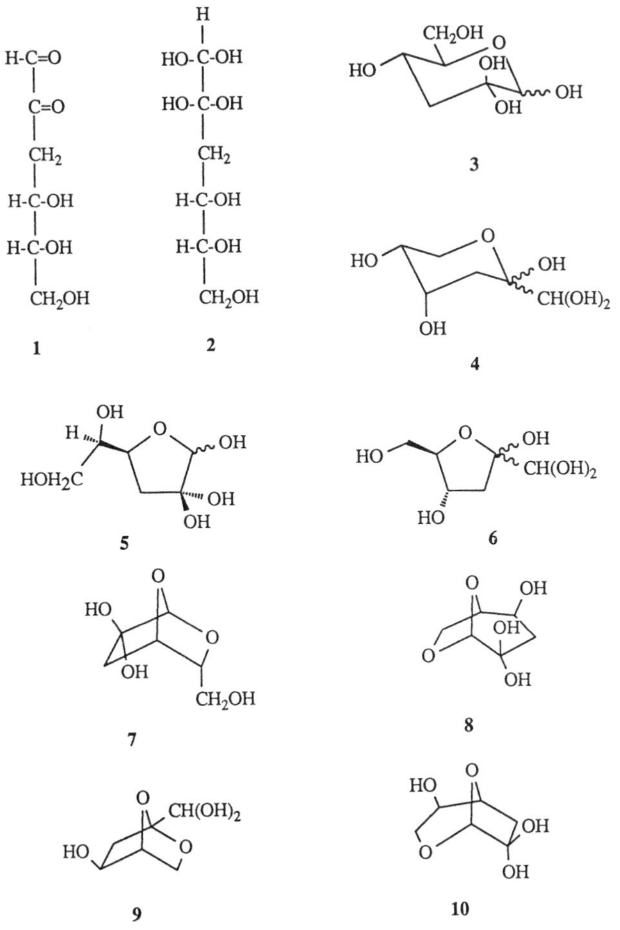

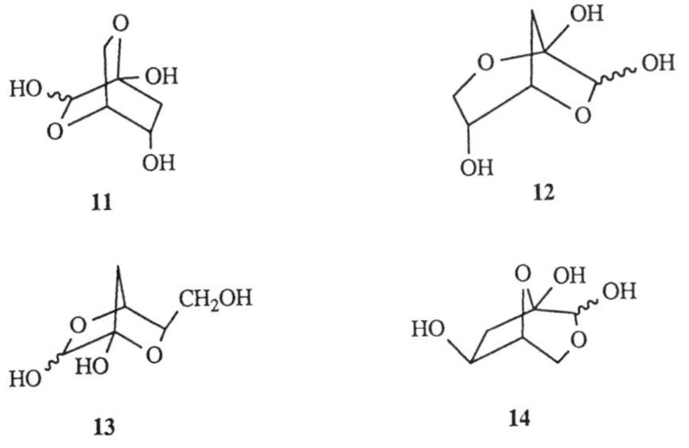

Scheme 1b. Hypothetical bicyclic hemi-acetal and hemi-ketal structures of 3-deoxyglucosone

adjacent to an electron withdrawing group. Scheme 1b shows the structures of the non-hydrated bicyclic hemi-acetal/hemi-ketal isomers (**11-14**). The bicyclic structures **11-14** are in agreement with the observed stability of 3-deoxyglucosone under thermal and acidic conditions (vide infra). DEPT, COSY and J-resolved ^{1}H-NMR of all major isomers were in complete agreement with the general structural features of 3-deoxyglucosone.

Heating a D_2O solution of 3-deoxyglucosone to 70° for 2 hours did not show H-D exchange, nor any other significant change in the ^{1}H-NMR and ^{13}C-NMR spectra. After storage of this osone for several months, the only sign of decomposition observed was the formation of small amounts of HMF. Apparently an aqueous solution of 3-deoxyglucosone is very stable, its conversion to a form with a free carbonyl at C-2 very slow, and the equilibrium of various forms not significantly affected by temperature variation between 30-70°.

NMR of 1-^{13}C-3-deoxyglucosone. 1-^{13}C-3-deoxyglucosone was synthesized from 1-^{13}C-glucose by the method of Khadem et al. (6). The ^{1}H-NMR and ^{13}C-NMR spectra clearly showed C-1 / H-1 couplings, but also long range C-1 / C-3 and C-1 / H-3 couplings. The ^{13}C-NMR (Fig. 3) confirmed the presence of 5 major and at least 6 minor isomers. A more detailed study of the structures of the isomers of 3-deoxyglucosone based on NMR will be published elsewhere.

Reactivity of 3-deoxyglucosone (Scheme 2)

Reaction with acid. 3-Deoxyglucosone is relatively stable under acidic conditions. When heated to 60° for 1 hour in 1 N CF_3COOD, less than 10% is converted to HMF. When a solution of 3-deoxyglucosone in aqueous acetic acid (2N) is heated to 100° for 1 hour, about 40-50% of the osone is converted to HMF. This is in agreement with what Anet has reported on 3-deoxyglucosone obtained from difructoseglycine (3). When the reaction is carried out in D_2O, about 20-30% D-incorporation is observed at the C-3 position of HMF. When heated to 100° for one hour in 1N CF_3COOD/D_2O complete decomposition of 3-deoxyglucosone is observed by NMR, and HMF and formic acid were found as reaction products. When we did the same reaction with 1-^{13}C-3-deoxyglucosone, the formic acid C-atom and the aldehyde C-atom of HMF were found to originate exclusively from C-1 of 3-deoxyglucosone. We assume levulinic acid (4-oxopentanoic acid) is formed as well, which together with formic acid has been reported to be formed from HMF under acidic conditions (16). The percentage D-incorporation in CD_3COOD/D_2O and CF_3COOD/D_2O was not significantly different. We had expected D-incorporation at C-3 of HMF to be much greater in CD_3COOD/D_2O, based on a report by Feather and Russell (17). These authors observed that when the 1-amino-1-deoxy-D-fructose derivatives derived from p-toluidine, dibenzylamine and morpholine are reacted with 2N acetic acid in D_2O, deuterium incorporation at C-3 in HMF ranges from 75-87%. When the same 1-amino-1-deoxy-D-fructose derivatives were reacted with HCl in D_2O, no D-incorporation took place. The formation of HMF from 1-amino-1-deoxy-D-fructose derivatives is expected to go via an enolic form of 3-deoxyglucosone. The authors conclude that in the HCl catalyzed reaction of 1-deoxy-1-amino-D-fructose derivatives to HMF, 3-deoxyglucosone is not a necessary intermediate. Our results are in agreement with their conclusion. Nevertheless, our observation that in the acid catalyzed conversion of

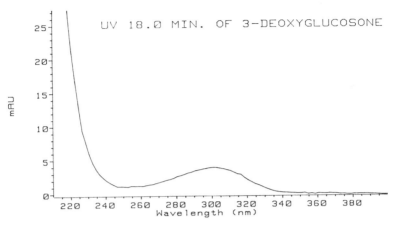

Figure 2. UV spectrum of 3-deoxyglucosone   (first peak in 220 nm chromatogram)

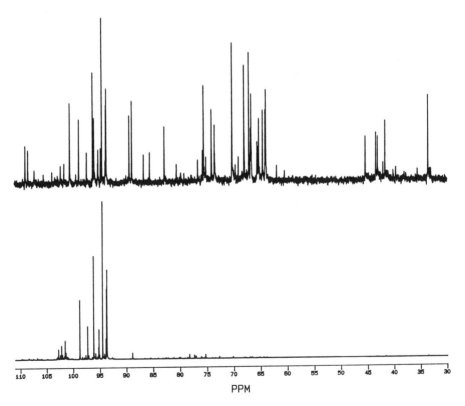

Figure 3. ^{13}C-NMR spectra of 3-deoxyglucosone and 1-^{13}C-3-deoxyglucosone

3-deoxyglucosone to HMF only 20-30% D is incorporated at C-3, is rather unexpected. We conclude that in the formation of HMF from 3-deoxyglucosone, either the free C-2 carbonyl is not a necessary intermediate, or the enol-keto tautomeric rearrangement is slow compared to the competing condensation reaction, which eventually leads to HMF. A mechanism for the formation of HMF from 3-deoxyglucosone not involving the free carbonyl at C-2 is outlined in scheme 3. The mechanism includes a furylium ion intermediate, which was also suggested by Yaylayan, to account for the formation of HMF from Amadori rearrangement products (18). We propose a similar mechanism for the bicyclic isomers 9, 13 and 14, which all contain a C-2/C-5 furanose ring.

Reaction with base. 3-Deoxyglucosone reacts quickly with aqueous base such as 0.1 M NaOH or 0.02 M Ca(OH)$_2$ to form metasaccharinic acid, in agreement with observations by Anet (3). However, HPLC analysis also showed some additional, as yet unidentified decomposition products, which are expected to include the oxidative cleavage product 2-deoxy-D-erythropentonic acid, as observed by Rowell and Green (19). When 3-deoxyglucosone was dissolved in 0.1 M K$_2$DPO$_4$ in D$_2$O (pH8), H-D exchange at C-3 was found to be complete after about 4 hours at ambient temperature. When 3-deoxyglucosone was heated to 130° for 1 hour in 0.1 M K$_2$HPO$_4$ in 1,2-propanediol (1% H$_2$O), all 3-deoxyglucosone had decomposed, and about 35-40 % metasaccharinic acid was formed.

Formation of pyrazines. The most important aspect of the reactivity of 3-deoxyglucosone in our opinion, is its cleavage to the smaller fragments, which are the building blocks of many flavour molecules such as pyrazines, thiazoles, carbocyclic compounds etc. We have used the formation of methylated pyrazines to study the fragmentation of 3-deoxyglucosone, and hexoses such as glucose and fructose in the Maillard reaction. We have tried to address the following questions regarding the formation of pyrazines:
1. How are pyrazines formed from the presumed dihydropyrazine intermediates?
2. How can the product distribution of the methylated pyrazines be explained?
3. What is the ratio of the 1- and 3-deoxyglucosone intermediates in the asparagine mediated pyrazine formation reaction?

Oxidation step in pyrazine formation. According to the generally accepted mechanisms for pyrazine formation in the Maillard reaction, pyrazines are formed in an oxidation step from their dihydropyrazine precursors. Since the presence or absence of oxygen did not affect the yield or product distribution of the formation of methylated pyrazines from fructose and asparagine, we rule out O$_2$ as the oxidizing species responsible for the conversion of dihydropyrazines to pyrazines. The fact that we do not find any tetrahydropyrazines or piperazines, also makes disproportionation very unlikely. This is in agreement with similar observations by Shibamoto (20). Major sugar breakdown products, however, are α-dicarbonyl species such as pyruvaldehyde, which would also be quite good H-acceptors. Similar to the disproportionation mechanism, pyrazines could be formed from dihydropyrazines in a redox reaction, in which a dicarbonyl species, e.g. pyruvaldehyde, is reduced to a hydroxycarbonyl species e.g. acetol. An indication of the validity of such a mechanism (scheme 4) is the fact that acetol (hydroxyacetone) has been observed in the headspace of the reaction of asparagine with fructose (H. Turksma, personal communication).

Scheme 2. Reactions of 3-deoxyglucosone

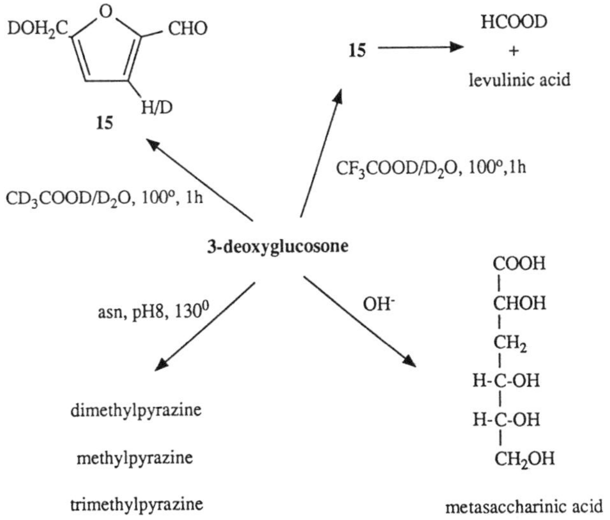

$15 \longrightarrow$ HCOOD + levulinic acid

CF_3COOD/D_2O, 100°, 1h

CD_3COOD/D_2O, 100°, 1h

3-deoxyglucosone

asn, pH8, 130°

OH⁻

dimethylpyrazine

methylpyrazine

trimethylpyrazine

COOH
|
CHOH
|
CH₂
|
H-C-OH
|
H-C-OH
|
CH₂OH

metasaccharinic acid

Scheme 3. Proposed mechansim for conversion of 3-deoxyglucosone to HMF

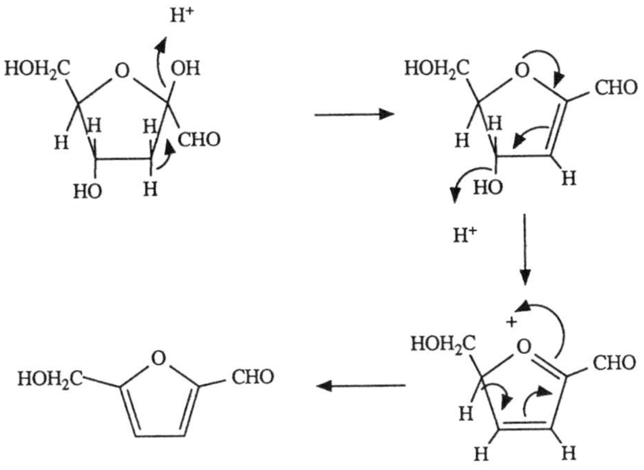

Scheme 4. Proposed mechanism for dihydropyrazine oxidation

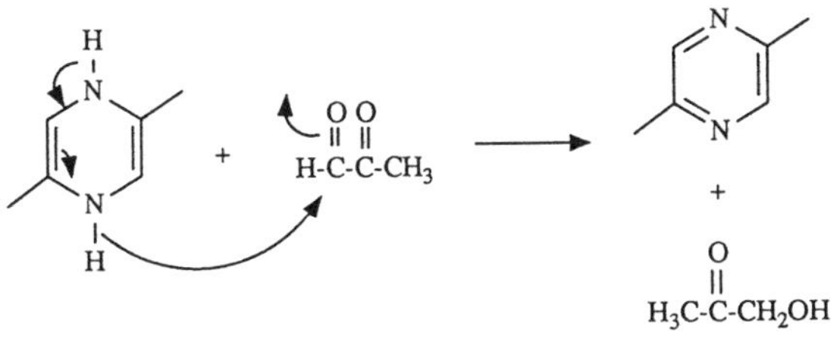

Another mechanism by which methylated (or alkylated) pyrazines can be formed, is dehydration of a hydroxymethylated (or hydroxyalkylated) dihydropyrazine (Scheme 7), as proposed by Shibamoto (20). It seems reasonable to assume that both mechanisms play a role in pyrazine formation.

Formation of methylated pyrazines. The question of the distribution of the methylated pyrazines has been investigated by the experiments indicated in table 1. When glucose or fructose and asparagine (asn) are heated in a solution with low water activity, a mixture of methyl pyrazine, dimethylpyrazine (predominantly 2,5- and 2,6-) and trimethylpyrazine is formed. Very small amounts (<0.05%) of other pyrazines such as 2,3-dimethylpyrazine, and tetramethylpyrazine are generated as well. Our working hypothesis was that glucose and fructose fragmentation takes place by retro-aldolization of the intermediate glucosones, to form mono- and α-dicarbonyl species (scheme 5). If this hypothesis is correct, pyruvaldehyde and 1-hydroxybutanedione should be the only α-dicarbonyl species generated. These could be expected to react further by Strecker degradation (scheme 6), condensation and oxidation, to form mainly dimethylpyrazine, but also trimethylpyrazine, and to a lesser extent tetramethylpyrazine. Dimethylpyrazine was indeed the main pyrazine formed, but unexpectedly methylpyrazine was the second largest product. When a solution of 3-deoxyglucosone and asparagine is heated, again a mixture of methyl-, dimethyl- and trimethylpyrazine is formed.

Koehler et al. (21) studied the formation of mono- and dimethylpyrazines from ^{14}C-glucose labeled at C-1, C-6 and both at C-3 and C-4. Their results indicate that dimethylpyrazine is formed from C-1/C-3 fragments as well as C-4/C-6 fragments, to about equal extent.

To account for our results we should therefore also consider the C-4/C-6 monocarbonyl species formed by retro-aldolization of the hexoses. In the case of pyrazine formation from 3-deoxyglucosone, glyceraldehyde is expected to be formed in addition to pyruvaldehyde (Scheme 5b). Glyceraldehyde can react with ammonia, generated by hydrolysis of asparagine, to form 1-amino-3-hydroxy-2-propanone. This aminoketone can combine with 1-aminopropanone, derived from pyruvaldehyde, to form dimethylpyrazine after losing three molecules of H_2O (Scheme 7). Glyceraldehyde can also undergo retro-aldolization to form hydroxyacetaldehyde and formaldehyde, which can give rise to methylpyrazine and trimethylpyrazine respectively (Scheme 7). Nonmethylated pyrazine is also a potential reaction product in this sequence, but is apparently not formed to a significant extent.

To account for pyrazines formed from glucose and fructose, fragmentation of 1-deoxyglucosone should also be considered. Retro-aldolization of 1-deoxyglucosone should proceed in a way which is similar to 3-deoxyglucosone retro-aldolization (Scheme 5), except that a four carbon fragment (hydroxybutadione) should also form. This hydroxybutadione can fragment further to pyruvaldehyde and formaldehyde, but it can also undergo Strecker degradation, and react with aminopropanone to give ultimately trimethylpyrazine. Fructose indeed seems to give a higher proportion of trimethylpyrazine than 3-deoxyglucosone.

Pyrazines from 1-^{13}C-glucose. To check our proposed mechanisms for asparagine mediated hexose fragmentation and pyrazine formation, and to determine the ratio of 1- versus 3-deoxyglucosone as intermediates in the

Table 1. Formation of pyrazines from hexoses and asparagine)[1]

C-source	N-source	Pyr)[2]	MMP	DMP	TrMP
fructose	Asn	-	2.3	4.9	0.4
glucose	Asn	-	1.8	2.6	0.1
3-Done	Asn	-	0.9	2.8	0.1

)[1] The yields of the reactions in this table were not optimized

)[2] Abreviations used:

Pyr = pyrazine
MMP = methylpyrazine
DMP = dimethylpyrazine
TrMP = trimethylpyrazine
Asn = asparagine

Scheme 5a. Primary products of retro aldol scission of 1-deoxyglucosone

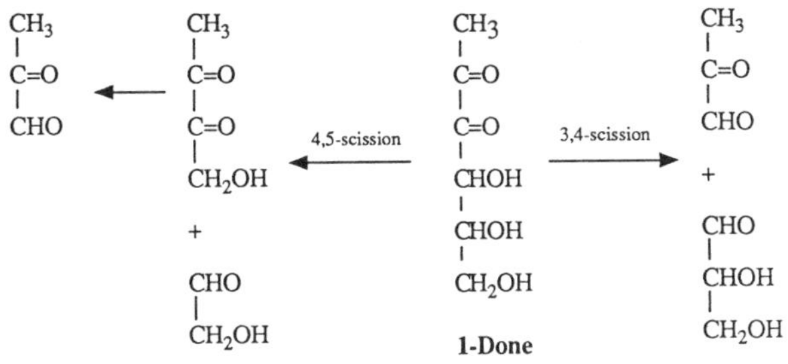

Scheme 5b. Primary products of retro aldol scission of 3-deoxyglucosone

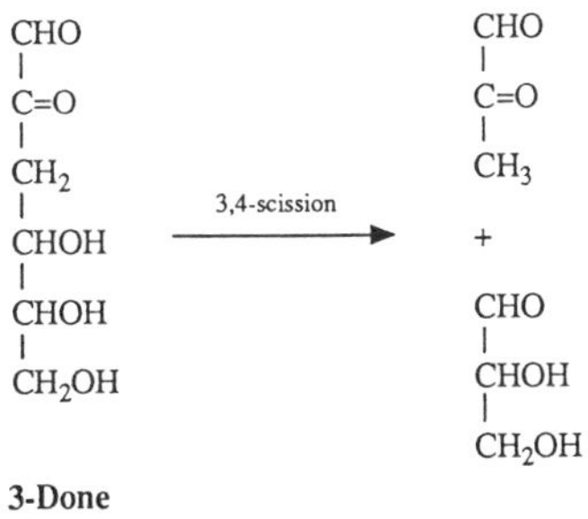

Scheme 6. Formation of dimethylpyrazine from pyruvaldehyde

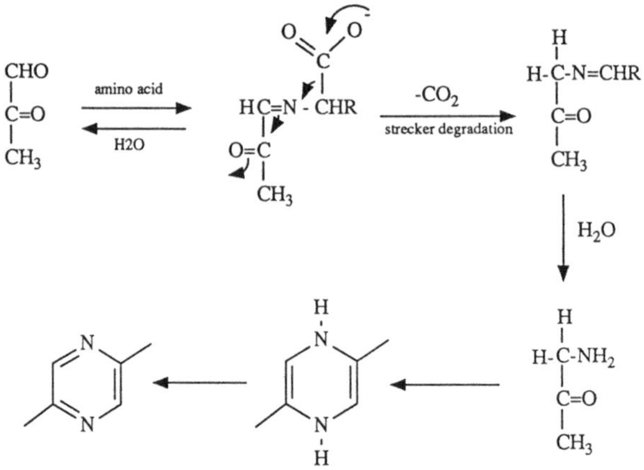

formation of methylated pyrazines, 1-^{13}C-glucose was reacted with asparagine. The pyrazines formed were isolated by Likens Nickerson extraction, and analysed by ^{1}H-NMR. C-H coupling was used to determine the position of the ^{13}C-label. In the major product, dimethylpyrazine, the methyl C-atoms contained about 26 % ^{13}C, and of the ring H-resonances (of all pyrazines combined) about 20 % showed ^{13}C-coupling. This suggests that 26 % of the C_3 fragments in dimethylpyrazine originate from pyruvaldehyde derived from 1-deoxyglucosone, and approximately 20 % from 3-deoxyglucosone (scheme 8), which is in general agreement with our proposed mechanisms (Scheme 7 & 8). The ^{13}C-NMR spectrum showed enriched ^{13}C-resonances for the methyl and methine carbons of the labelled pyrazines, and, as expected, no enriched ^{13}C-resonances for the quarternary C-atoms.

It must be realized that intermediates other than 1- and 3-deoxyglucosone, and cleavage mechanisms other than retro-aldolization cannot be excluded at this point. A report on Strecker degradation products from 1-^{13}C-D-glucose and glycine suggests that osones other than 1- and 3-deoxyglycosones occur in the Maillard Reaction (8). Moreover, Ledl and Schleicher (10) are suggesting that in addition to retroaldolization, glycosones are cleaved oxidatively between the two carbonyls. 3-Deoxyglucosone was observed to cleave to 2-deoxypentonic acid/lactone and formic acid. The 2-deoxypentonic acid lactone can be expected to undergo further fragmentation (by retroaldolization) to fragments which could act as pyrazine precursors, thus explaining the loss of ^{13}C-label in the methylated pyrazines, as observed by us. A similar reaction sequence could occur via 1-deoxyglucosone. We are presently investigating the nature of the fragmentation processes in the Maillard Reaction in more detail.

Main conclusions

1. In aqueous solution 3-deoxyglucosone occurs as 5 major and at least 6 minor isomers.

Scheme 7. Formation of pyrazines from 3-deoxyglucosone (3-Done) and asparagine (asn)

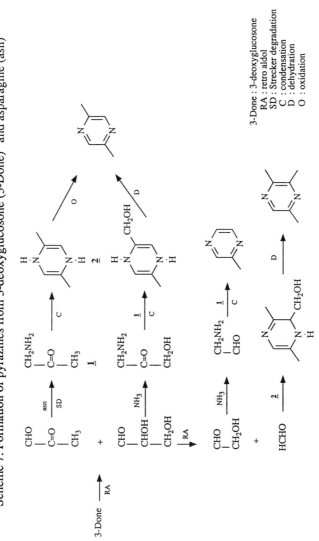

3-Done : 3-deoxyglucosone
RA : retro aldol
SD : Strecker degradation
C : condensation
D : dehydration
O : oxidation

Scheme 8. Formation of ^{13}C-labeled fragments from 1-^{13}C-D-glucose by retro-aldolization

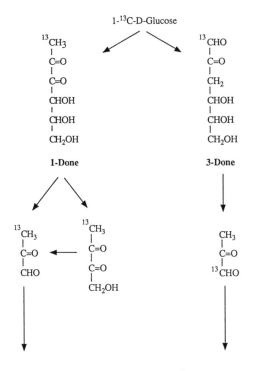

2. Absence of H-D exchange indicates that under neutral conditions isomerization to the free keto form of 3-deoxyglucosone is insignificant.
3. 3-Deoxyglucosone is relatively stable under acidic conditions, again showing remarkably little H-D exchange at C-3.
4. At pH8 H-D exchange at C-3 is much faster, and complete in about 4 hours.
5. At pH8 glucose reacts with asparagine to give methylated pyrazines. The fragmentation process underlying the formation of the methylated pyrazines can be understood assuming retro-aldolization of 1- and 3-deoxyglucosone.

Acknowledgments We thank Dr. H. Turksma for his valuable suggestions, and H.J.J. van Heck, W.L. Koelewijn, R. Plomp and H. Vonk for their technical assistance.

Literature cited

1.   Ishizu, A.; Lindberg, B.; Theander, O. *Carbohydr. Res.* **1967,** *5(3),* 329-324.
2.   Anet, E.F.L.J. *Austr. J. Chem.* **1960,** *13,* 396-403.

3. Anet, E.F.L.J. *J. Am. Chem. Soc.* **1960**, *82*, 1502.
4. Kato, H. *Bull. Agr. Chem. Soc. Japan* **1960**, *24*, 1-12.
5. Kato, H., Hayase, F., Shin, D.B., Oimoni, M., Baba, S. In *The Maillard Reaction in Aging, Diabetes and Nutrition;* Baynes, J.W., Monnier,V.M., Eds.; Progr. Clin. Biol. Res. 304; A.R. Liss Inc., New York, 1989, pp 69-84.
6. Khadem, H.El.; Horton, D.; Meshreki, M.H., Nashed, M.A. *Carbohydr. Res.* **1971**, *17*, 183-192.
7. Madson, M.A.; Feather, M.S. *Carbohydr. Res.* **1981**, *94*, 183-191.
8. Nyhammar, T.; Olsson, K.; Pernemalm, P.A. In *The Maillard Reaction in Foods and Nutrition;* Waller, G.R.; Feather, M.S., Eds.; ACS symposium series 215; Am. Chem. Soc., Washington, D.C., 1983; pp 71-82.
9. Beck, J.; Ledl, F.; Severin, T. *Z. Lebensm. Unters. Forsch.* **1989**, *188*, 118-121.
10. Ledl, F., Schleichr, E. *Angew. Chemie Int. Ed. Engl.* **1990**, *29*, 565-594.
11. Anet, E.F.L.J. *Advan. Carbohydr. Chem.* **1964**, *19*, 181-218.
12. Baltes, W.; Kunert-Kirchhoff, J.; Reese, G. In *Thermal generation of aromas;* Parliment, T.H.; McGorrin, R.J.; Ho, C.-T., Eds.; ACS symposium series 409; Am. Chem. Soc., Washington, D.C., 1989; pp 143-155.
13. Tressl, R.; Helak, B.; Martin, N.; Kersten, E. In *Thermal generation of aromas;* Parliment, T.H.; McGorrin, R.J.; Ho, C.-T., Eds.; ACS symposium series 409; Am. Chem. Soc., Washington, D.C., 1989; pp 156-171.
14. Corbett, W.M. In *Methods in Carbohydrate Chemistry;* Whistler, R.L.; Wolfrom, M.L.; BeMiller, J.N., Eds.; Vol. II; Academic Press Inc., New York, London, 1963, pp 480-482.
15. Godefroot, M.; Sandra, P.; Verzele, M; *J. Chromatogr.* **1981**, *203*, 325-335.
16. Pigman, W.; Anet, E.F.L.J. In *The Carbohydrates;* Pigman, W.; Horton, D., Eds.; Academic Press, New York, London, San Fransisco, 1972; p 185.
17. Feather, M.S.; Russell, K. *J. Org. Chem.* **1969**, *34(9)*, 2650-2652.
18. Yaylayan, V. *Trends Food Sci. Tech.* **1990**, *july*, 20-22.
19. Rowell, R.M., Green, J. *Carbohydr. Res.* **1970**, *15*, 197-203.
20. Shibamoto, T., Bernhard, R.A., *Agric. Biol. Chem.,* **1977**, *41*, 143-153.
21. Koehler, P.E., Mason, M.E., Newell, J.A., *J. Agr. Food Chem.,* **1969**, *17*, 393-396.

RECEIVED January 30, 1992

Chapter 18

Formation of Smoke Flavor Compounds by Thermal Lignin Degradation

Reiner Wittkowski[1], Joachim Ruther[2], Heike Drinda[2], and Foroozan Rafiei-Taghanaki[2]

[1]Max von Pettenkofer-Institut, Bundesgesundheitsamt, Postfach 33 00 13, D–1000 Berlin 33, Germany
[2]Institut für Lebensmittelchemie, Technische Universität Berlin, Gustav-Meyer-Allee 25, D–1000 Berlin 65, Germany

Phenolic compounds as most important flavor compounds arise from oxidative lignin decomposition. Lignin units exhibiting guaiacol structures are preformed in softwood lignin but are also formed from syringol units in hardwood lignin by demethylation of one methoxy substituent. Pyrocatechols and alkylated phenols as secondary pyrolysis products are formed by radical cleavages and recombinations. Headspace enrichment procedures from smoked ham and subsequent sensory evaluation have proven that guaiacol and some of its alkyl isomers are the dominating smoke flavor compounds in smoked foods. This is true for the products smoked with both softwood and hardwood.

Some thousand years ago men developed smoke treatments in his efforts to inhibit spoilage and to enhance the keeping qualities of foodstuffs, meat and fish in particular. In Europe smoked foods are still very popular and also in the United States smoke flavor is used in several meat products, barbecue sauces and is used for other flavoring purposes. The main reason for using smoke treatments today, however, is to impart a smoke flavor to the finished product and to yield a certain smoke color and not to exploit the preserving properties.

Three reasons were decisive for focussing the analytical interest to the phenolic compounds in curing smoke. Firstly, phenols are associated with almost all of the desired technological effects of smoking: coloring, preserving and flavor formation. Secondly, they become strongly enriched when condensing smoke in an aqueous solution. The third reason is due to the reported carcinogenic or cocarcinogenic properties of some phenols (1-4).

0097–6156/92/0490–0232$06.00/0
© 1992 American Chemical Society

Smoke flavor

Considerable work has been done during the last decade (5-10) to isolate and identify phenolic compounds from curing smoke. Finally, 211 compounds were identified in the phenolic fractions of smoke condensates, 156 phenols, 33 other aromatic alcohols and 22 hydroxylated heterocycles as well as 23 phenolic acids (6). Their analysis was performed by use of preparative clean-up procedures such as column chromatography on silylated silica, aluminium oxide and HPLC on RP-18 in combination with high resolution GC and mass spectrometry including different ionization modes (EI,CI) and a special accurate mass measurement at low resolution, which gave rise to the sum formulas of all measured fragments detected in one gas chromatographic run (11, 12). The results led to a comprehensive elucidation of the fragmentation pathways of silylated phenols. These results along with detailed descriptions on the analysis of phenols in curing smoke have been published (6).

An overview of the basic phenol structures is given in Figure 1. All theoretical combinations for hydrogen and methyl substituents were found for R1, R2, and R3 by reference (6). The upper structure includes syringol, 3-methoxypyrocatechol, pyrogallol, and 2,3-dimethoxyphenol. 1,2,3-Trimethoxybenzene, a non-phenolic compound, was also identified and fits to this general structure. Phenols, guaiacols, pyrocatechols, resorcinols, and 3-methoxyphenols are deducible from this basic formula. The substituent R4 is predominantly situated in the para position and consists of saturated and unsaturated alkyl side chains with up to three carbon atoms. Some phenols possess functional groups in their side chains.

In some cases the smoke flavor has been attributed to the presence of phenols (13-15) and sometimes to a single component. Attempts to duplicate the flavor of smoked meat products with various mixtures of phenols or single components, even with phenolic fractions isolated from curing smoke condensates have been unsuccessful. This is based on the fact that the flavor of smoke condensates on the one hand and the flavor of smoked ham, bacon, and sausages on the other hand (5) can be clearly distinguished. Though the phenols of smoke condensates and smoked foods are well studied their role and extent during flavor formation was unclear for a long time. It could be confirmed, however, that an array of certain phenols was necessary to produce a desirable flavor in smoked meat products (16).

Recent investigations have clarified the extent of contribution of smoke components to the flavor of smoked ham. The volatile compounds have been analyzed by GC-MS after isolation from the headspace of those foodstuffs by static (17) and dynamic sampling procedures (18). The static procedure with sampling and injection volumes of up to 20 ml of headspace produced chromatograms exhibiting high concentrations of terpenes. Their contribution to the smoke flavor, however, is very small. The flavor compounds of interest could be isolated by dynamic headspace enrichment. Figure 2 shows a gas chromatogram obtained from the headspace of traditionally smoked ham. The flavor compounds had been stripped from the headspace for 48 hours and subsequently trapped on Tenax (18). After GC separation the effluent was split and the volatiles were sensorially evaluated by a test panel using the sniffing technique (19, 20). The indicated peaks in Figure 2 point to those components which were associated with smoke-type aroma impressions by the test

Substance	R_1	R_2	R_3
Syringol	H	CH_3	CH_3
3-Methoxypyrocatechol	H	CH_3	H
2-Methoxyresorcinol	CH_3	H	H
Pyrogallol	H	H	H
2,3-Dimethoxyphenol	CH_3	CH_3	H

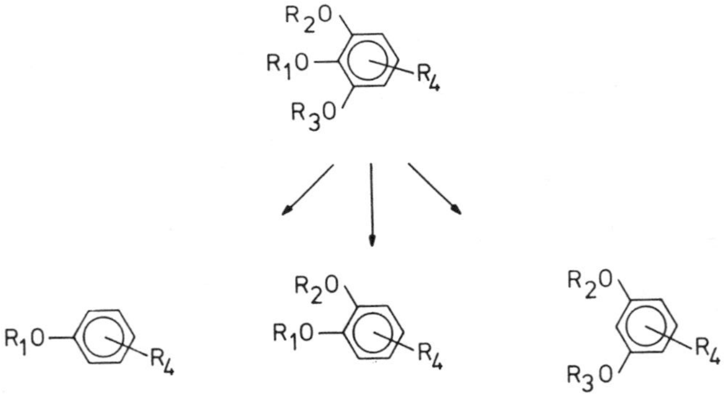

Substance	R_1	R_2
Resorcinol	H	H
3-Methoxyphenol	CH_3	H

Substance	R_1	R_2
Guaiacol	H	CH_3
Pyrocatechol	H	H

Substance	R_1
Phenol	H

Figure 1: Overview of the basic phenolic structures found in curing smoke (see text for details)

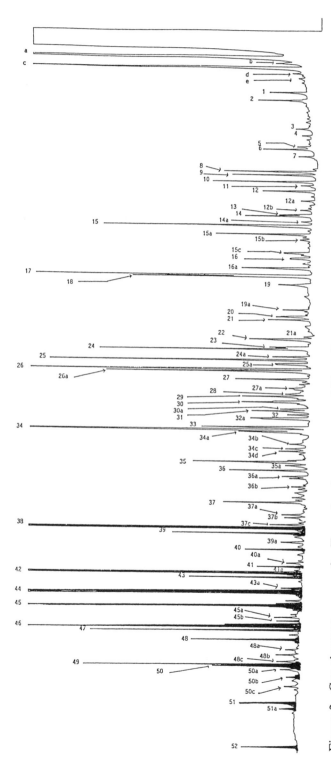

Figure 2. Gas chromatograms of a Tenax headspace extract of Black Forest Ham. Column: 60 m × 0.32 mm DB-Wax (J&W), film thickness 0.5 μm. The indicated peaks were associated with smoke flavor notes: 38: guaiacol; 39: 6-methylguaiacol; 40: 2,6-dimethylphenol; 41: 5-methylguaiacol; 42: 4-methylguaiacol; 43: C2-subst. guaiacol; 43a: 2,4,6-trimethylphenol; 44: o-cresol + phenol; 45: 4-ethylguaiacol; 46 I: 2,5-dimethylphenol; 46 II: p-cresol + 2,4-dimethylphenol; 47: m-cresol; 48: 4-propylguaiacol; 49: eugenol; 50: 4-ethylphenol; 50b: 4-vinylguaiacol; 51: isoeugenol (*cis*); 51a: 4-propylphenol; 52: isoeugenol (*trans*).

panel. The same procedure was applied to ham treated with hardwood smoke to clarify the role of syringol and its derivatives in regard to their flavor contribution in the final product (21). The aroma importance of each compound was determined by the flavor dilution method developed by Grosch (22). This involves a stepwise dilution of the extract and a subsequent sensory evaluation by GC effluent sniffing by a test panel after each dilution step. This offers the great advantage that not only the concentrations of specific components are measured but in parallel their aroma intensity and impression. Hence, the resultant flavor dilution factor describes the aroma importance of a compound in that specific aroma extract. It could be demonstrated as a major result that guaiacol, 4-, 5-, and 6-methylguaiacol and 2,6-dimethylphenol are the dominating flavor compounds in the headspace extracts of smoked ham. Surprisingly, this is true both for ham smoked with softwood and ham smoked with hardwood. Syringol and its derivatives play only a minor role in this view, probably because of their small vapor pressure. After thirty years this is an analytical result supporting the empirical findings of Spanyar et al. (23) and Tilgner (24) who found that the smoke from both softwood and hardwood yield smoke flavors of comparable quality.

In a recent study the most important phenols have been submitted to sensory evaluation to correlate the aroma impression to the concentration (25). Here especially guaiacol and some of its alkyl derivatives exhibited a remarkable increase of smoked ham flavor impression with increasing concentration. Figure 3 exemplarily shows this effect. A test panel of eight members tested different concentrations of 4-methylguaiacol in aqueous solutions. At concentrations below 0.01 ppm a smoky aroma note was only determined by one participant, whereas five of them described the flavor as musty. At a concentration level of 0.1 ppm all members described the solution as smoky. A caramel-like flavor note was also recognized.

Lignin

The phenolic components in curing smoke originate mainly from the pyrolysis of lignin, which is one of the major constituents of wood (about 25%). In biosynthesis lignin arises from a free-radical copolymerization of three phenylpropanoid monomers, namely coumaryl-, coniferyl-, and sinapyl alcohol leading to a three-dimensional all round ramified macromolecule (26). Based on a series of investigations a constitution scheme of beechwood lignin was developed by Nimz (27) which is shown in Figure 4. This implies, however, that a model scheme can only reflect to the statistical distribution of structural units, which had been confirmed by analytical investigations (26). Nevertheless, the prior existence of the guaiacol- and syringol propane units connected via ether bindings is apparent.

Formation of Phenols by Lignin Pyrolysis

Under the topic "biomass thermolysis and conversion" considerable work has been reported concerning the elucidation of phenol formation during wood combustion. In

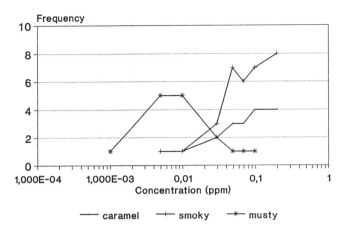

Figure 3: Dependence of the odor character of 4-methylguaiacol on its concentration in an aqueous solution (see text for details).

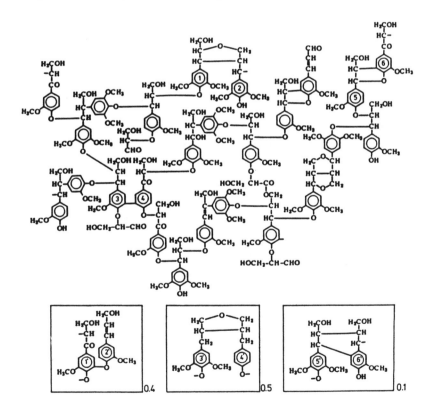

Figure 4: Constitution scheme of beechwood lignin (27). The dilignols 1'2', 3'4', and 5'6' can be exchanged against the corresponding units in the macromolecule with the indicated frequency factors. (Adapted from ref. 6).

general, several model investigations have been performed, where lignin monomers and related compounds have been pyrolyzed to help elucidate the fragmentation processes (28-33). Fiddler et al. (29) and Kossa (32) found that heating ferulic acid at temperatures of 240 to 260 °C was sufficient to effect a stepwise propanoid side chain degradation leading to the final product guaiacol. Hurff and Klein (30) investigated the hydrodeoxygenation of guaiacol and anisole over a CoO-MoO$_3$/Al$_2$O$_3$ hydrotreated catalyst and obtained the demethylation products phenol and pyrocatechol, respectively. Petrocelli and Klein (33) emphasize the secondary decomposition reaction of guaiacols to pyrocatechols at 400 °C. In the range of 390 to 420 °C a strong exothermic reaction takes place (34), which Wienhaus (35) explains as the recombination of free radicals. Connors et al. (28) refer these reactions to the reactive methoxy group of guaiacols and syringols. By pyrolysis of 4-ethylguaiacol they observed cleavages of the O-C (alkyl) and O-C (aryl) bonds to give 4-ethylphenol and 4-ethylpyrocatechol as main products. Moreover, 2-ethylphenol arose by rearrangement in the mesomeric stable phenyl radical. Guaiacol and its 4-alkyl derivatives with up to three carbon atoms are predominant constituents of the phenolic fraction of curing smoke. Some other pyrolytic fragments contain functional groups in their side chains, for example phenylalcohols, aldehydes and ketones.

Our experiments have shown that pyrocatechol and a certain number of its derivatives are found in relatively high amounts in curing smoke (5-10). According to Marton (36) lignin contains only one catecholic aromatic unit, which could lead directly to the occurrence of catecholic fragments by simple side chain degradation. The fact that the predominant derivatives of guaiacol and pyrocatechol show similarities in their substituent structure supports the hypothesis that O-C cleavage in the guaiacol ortho methoxy group is responsible for the second hydroxy group formation, according to the authors mentioned above.

Along with pyrocatechols high yields of alkyl- and dialkylphenols are found in curing smoke condensates. All types of dimethylphenols and some methylethylphenols are main constituents of the phenolic fraction (5-10, 37).

To provide information on the reaction types and yields of pyrolytic fragments at temperatures higher than 400 °C and to allow one general reaction scheme to phenol formation pathways in this regard a pyrolysis of ferulic acid under conditions comparable to oxidative wood combustion, namely temperatures higher than 400 °C and an abundant supply of oxygen was carried out (6). To illustrate the decomposition course of ferulic acid a sample was continuously heated and the sum of arising pyrolytic fragments recorded as FID signal. Figure 5 demonstrates the occurence of two maxima at 242 and 380 °C. In the range between 230 to 260 °C the main products found were degradation fragments formed by the destruction of the propanoid side chain. In addition to methyl-, ethyl-, and vinylguaiacol, vanillin and acetovanillone were identified. Moreover, isopropylguaiacol, eugenol, and isoeugenol were formed in relatively high amounts during pyrolysis, which indicates other reaction types than the stepwise side chain degradation. These data are in agreement with the results reported by Fiddler et al. (29) and Kossa (32). The highest yield of pyrolytic fragments, however, was obtained in the range between 360 and 410 °C. This represents the temperature range, where radical C-C cleavages and recombinations were previously observed (*34, 35, 38, 39*). Each condensate obtained by oxidative pyrolysis at temperatures of more than 400 °C showed pH values of 4.2. They were yellow (40% methanol in water) and light brown (water) colored. The aroma of the methanol solution was characterized as sooty and burnt. The aqueous solutions, however, were aromatic and mild smoky.

Mass spectrometric analysis of the trimethylsilyl derivatives separated by GC permitted the identification of 58 phenolic compounds (*6*), see Table I. Besides the pyrolytic changes in the side chain, alterations at the aromatic ring were apparent leading to the formation of three basic phenolic structures, namely phenol, guaiacol, and pyrocatechol, about 45% of which belonged to the pyrocatechol class. Guaiacol derivatives, however, occurred in a portion of less than 16%. Moreover, it was conspicuous that guaiacol itself and the homolog derivatives with functional groups containing side chains were absent, although they belong to the main reaction products at pyrolysis temperatures less than 300 °C. This leads to the consequence that in accordance with Petrocelli and Klein (*33*) pyrocatechols have to be regarded as secondary pyrolysis products, whereas guaiacols appear to be intermediates. Also noteworthy were the high amounts of dialkylphenols, particulary those with methyl groups in the ortho position (2-methyl-4-vinylphenol, 9.35%; 2-methyl-4-ethylphenol, 7.25%; 2,4-dimethylphenol, 2.67%). This points to rearrangement reactions in which the methoxy group is transformed to a methyl group.

Furthermore, the methyl- and dimethyl- as well as methylethyl- and vinylphenols originate from guaiacol intermediates. It is noteworthy that methyl alkylation in the ortho position to the phenolic hydroxy group is dominant. Based on the determination of the phenolic reaction products a postulated general reaction scheme is shown in Figure 6 in a simplified form. In reality, lignin degradation pathways are much more complex than illustrated in Figure 6. It does not take into account mass balances including other reaction prodcuts such as carbon dioxide, water, and the tar-condensate fraction to make allowance for a complete kinetics calculation. Also the role of oxygen and the extremely stable phenoxy radicals as possible important intermediates are not included.

Starting from the guaiacol structure of ferulic acid, the reaction pathway C describes the stepwise side chain degradation to form guaiacol derivatives. The loss of a methyl radical from the methoxy group leads to the formation of pyrocatechols and methane by recombination with two hydrogen radicals (A). The dialkylphenols derive from the reaction of a methyl with a phenyl radical, which originates from the elimination of the whole methoxy group. The methoxy radical combines with a hydrogen radical to form methanol (B).

The high amounts of ortho methylated phenols lead to the assumption that alkylations mainly or even completely arise from the addition of methyl groups in the position of the guaiacol methoxy group. Another possibility is the rearrangement of the side chain before recombination, where R2 is a second methyl radical arising from the methoxy cleavage. The recombination of free radicals is always an exothermic reaction (*34, 35*). Such an exothermic process was found at about 400 °C during lignin pyrolysis by Domburg et al. (*38, 39*). Fenner and Lephard (*40*) observed a maximum of methanol formation by application of FTIR spectroscopy in this same temperature range. The recombination of radically formed phenol fragments

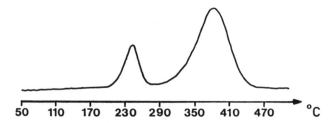

Figure 5: Thermal decomposition of ferulic acid. (Adapted from ref. 6).

Table I. Relative phenol composition of the condensate obtained by oxidative pyrolysis of ferulic acid

phenol		guaiacol		pyrocatechol	
R	%	R	%	R	%
-	1.2	4-CH=CH$_2$	1.2	-	7.5
2-CH$_3$	2.48	CH=CH$_2$?	0.83	3-CH$_3$	2.69
3-CH$_3$	0.23	CH=CH$_2$?	0.37	4-CH$_3$	5.84
4-CH$_3$	1.98	CH=CH$_2$?	trace	4-CH=CH$_2$	9.73
2-CH=CH$_2$	0.2	4-CH$_2$-CH$_3$	2.6	3-CH$_2$-CH$_3$	0.27
3-CH=CH$_2$	0.1	4-CH$_2$-CH=CH$_2$	0.08	4-CH$_2$-CH$_3$	10.73
4-CH=CH$_2$	1.88	4-CH$_2$-CH$_2$-CH$_3$	0.13	4-CH$_2$-CH=CH$_2$	0.58
2-CH$_2$-CH$_3$	0.3			4-CH=CH-CH$_3$ cis	0.55
3-CH$_2$-CH$_3$	0.25			4-CH=CH-CH$_3$ trans	1.58
4-CH$_2$-CH$_3$	3.45			-CH$_3$;CH=CH$_2$	0.19
2-CH$_3$;3-CH$_3$	0.25			4-CH$_2$-CH$_2$-CH$_3$	1.2
2-CH$_3$;4-CH$_3$	2.67			-C$_3$H$_7$	0.67
2-CH$_3$;5-CH$_3$	0.1			-C$_3$H$_7$	0.25
2-CH$_3$;6-CH$_3$	0.37			-C$_4$H$_9$	0.12
3-CH$_3$;4-CH$_3$	0.89			4-C$_6$H$_5$?	0.21
3-CH$_3$;5-CH$_3$	0.06			4-CHO	0.18
4-CH$_2$-CH=CH$_2$	0.14			4-CO-CH$_3$	1.81
2-CH$_3$;4-CH=CH$_2$	9.35			4-CO-CH$_2$-CH$_3$	0.8
-CH$_3$;-CH=CH$_2$	0.12			4-CH=CH-COOH	1.0
-CH$_3$;-CH=CH$_2$	0.03				
-CH$_3$;-CH=CH$_2$	trace				
-CH$_3$;-CH$_2$-CH$_3$	trace				
-CH$_3$;-CH$_2$-CH$_3$	0.25				
-CH$_3$;-CH$_2$-CH$_3$	0.16				
-CH$_3$;-CH$_2$-CH$_3$	0.15				
-CH$_3$;-CH$_2$-CH$_3$	0.06				
2-CH$_3$;4-CH$_2$-CH$_3$	7.25				
hydroxybenzo-furan ?	0.15				
4-CHO	0.04				
2-CH=CH-COOH	0.04				
3-CH-CH-COOH	0.16				
4-CH=CH-COOH	0.13				
total	34.44		15.77		45.85

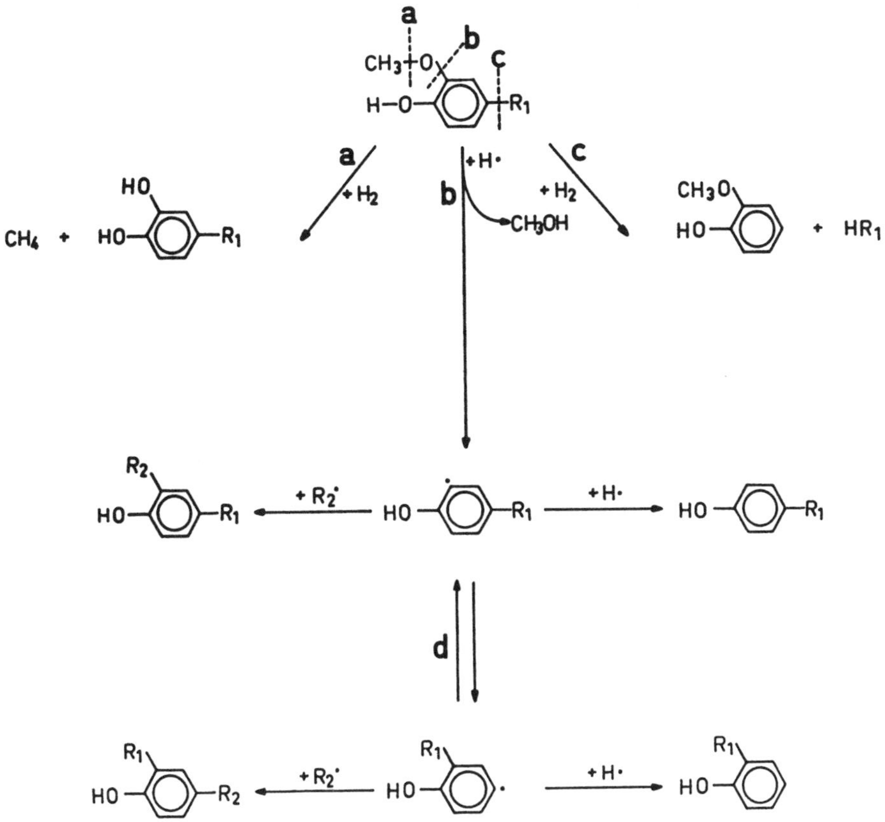

Figure 6: Degradation pathways in lignin pyrolysis and formation of phenols as secondary reaction products. (Adapted from ref. 6).

with methyl radicals is also interesting from another point of view. The findings of three carbon containing side chains in the para position point to homolytic cleavages also in the propanoid side chain, as indicated by Domburg et al. (*38, 39*). If recombinations with radicals serve as a basis, the formation of isopropyl-, propenyl-, and allylphenols is not surprising. The recombination with hydrogen radicals, however, develops the same products as those formed by heterolytic side chain cleavages.

In summary, it has been emphasized that the guaiacol derivatives which arise at about 250 to 300 °C in the course of lignin pyrolysis lead to the formation of pyrocatechols by transformation of the methoxy group into a hydroxy group at temperatures higher than 400 °C. Simultaneously, methylated products are developed, whereas the methyl radicals evidently arise from the O-C(alkyl) cleavage of the methoxy group. The exchange of the methoxy- with a methyl or hydrogen radical produce alkyl- and dialkylphenols. Therefore these substances must be regarded as stable final products formed by lignin pyrolysis. These data are in accordance with the investigations of Toth (*5*), who observed a relative decrease of guaiacol along with a relative increase of pyrocatechols at an increasing decomposition temperature from 600 to 1000 °C in the aqueous condensates of curing smoke.

In correlation with the recent findings concerning the concentration dependence of the smoke flavor properties of guaiacol and its alkyl derivatives in smoked ham it becomes evident why pyrolysis temperatures of wood in the smoking process should not exceed 600 °C.

Literature Cited

1.   Gibel, W.; Gummel, H. *Dtsch. Gesundheitswes.* **1967**, *22*, 980.
2.   Kaiser, H. E.; Bartone, K. C. *J. Nat. Med. Assoc.* **1966**, *58*, 361.
3.   Kaiser, H. E. *Cancer* **1967**, *20*, 614.
4.   Van Duuren, B. L.; Katz, C.; Goldschmidt, B. M. *J. Nat. Cancer Inst.* **1973** *51*, 703.
5.   Toth, L. *Chemie der Räucherung*; Verlag Chemie: Weinheim, 1982.
6.   Wittkowski, R. *Phenole im Räucherrauch -Nachweis und Identifizierung-*; VCH-Verlagsgesellschaft: Weinheim, 1985.
7.   Wittkowski, R.; Toth, L.; Baltes, W. *Lebensmittelchem. Gerichtl. Chem.* **1981**, *35*, 61.
8.   Wittkowski, R.; Toth, L.; Baltes, W. *Z. Lebensm. Unters.-Forsch.* **1981**, *173*, 445.
9.   Baltes, W.; Wittkowski, R.; Söchtig, I.; Block, H.; Toth, L. In *The Quality of Foods and Beverages, Chemistry and Technology*; Charalambous, G.; Inglett, G.; Eds.; Academic Press: New York and London, 1981, Vol. 2, pp 1-19.
10.   Toth, L.; Wittkowski, R.; Baltes, W. In *Recent Developments in Food Analysis*; Baltes, W.; Czedik-Eysenberg, P. B.; Pfannhauser, W.; Eds.; Verlag Chemie: Weinheim (1982), pp 70-75.

11. Wittkowski, R.; Kellert, M.; Baltes, W. *Int. J. Mass Spectrom. Ion Phys.* **1983**, *48*, 339.
12. Wittkowski, R.; Kellert, M.; Blaas, W.; Baltes, W. *Lebensmittelchem. Gerichtl. Chem.* **1984**, *38*, 69.
13. Baltes, W.; Söchtig, I. *Z. Lebensm. Unters.-Forsch.* **1979** , *169*, 9.
14. Wasserman, A. E. *J. Food Sci.* **1966**, *31*, 1005.
15. Fujimaki, M.; Kim, K.; Kurata, T. *Agr. Biol. Chem.*, **1974**, *38*, 45.
16. Toth, L.; Wittkowski, R. *Chemie in unserer Zeit* **1985**, *19*, 48.
17. Wittkowski, R.; Baltes, W.; Jennings, W. G. *Food Chemistry* **1990**, *37*, 135.
18. Wittkowski, R.; Ruther, J. in preparation 1991.
19. Rapp, A.; Knisper, W. *Chromatographia* **1980**, *13*, 698.
20. Rapp, A.; Knisper, W. *Vitis* **1980**, *19*, 13.
21. Wittkowski, R.; Rafiei-Taghanaki, F. in preparation 1991.
22. Grosch, W. *Chemie in unserer Zeit* **1990**, *24*, 82.
23. Spanyar, P.; Kevei, E.; Kiszel, M. *Z. Lebensm. Unters.-Forsch.* **1960**, *112*, 471.
24. Tilgner, D. J. *Fleischwirtsch.* **1958**, *10*, 751.
25. Wittkowski, R.; Drinda. H. in preparation 1991.
26. Krüger, G. *Chemie in unserer Zeit* **1976**, *10*, 21.
27. Nimz, H. *Angew. Chem.* **1974**, *86*, 336.
28. Connors, W. J.; Johanson, L. N.; Sarkanen, K. V.; Winslow, P. *Holzforschung* **1980**, *34*, 29.
29. Fiddler, W.; Parker, W. E.; Wasserman, A. E.; Doerr, R. C. *J. Agric. Food Chem.* **1967**, *15*, 757.
30. Hurff, S. J.; Klein, M. T. *Ind. Eng. Chem. Fundam.* **1983**, *22*, 426.
31. Klein, M. T.; Virk, P. S. *Ind. Eng. Chem. Fundam.* **1983**, *22*, 35.
32. Kossa, T. *Gaschromatographisch-massenspektrometrische Untersuchung flüchtiger und schwerflüchtiger Inhaltsstoffe von Bierwürze*; Dissertation, Technische Universität Berlin, D 83, 1976.
33. Petrocelli, F. P.; Klein, M. T. In *Fundamentals of Thermochemical Biomass Conversion*; Overend, R. P.; Milne, T. A.; Mudge, L. K., Eds.; Elsevier Applied Science Publishers: London and New York, 1985; pp 257.
34. Sarkanen, K. V.; Ludwig, C. H. In *Lignin*; Wiley: New York, 1971; pp 576 and 581.
35. Wienhaus, O. *Holztechnologie* **1979**, *20*, 144.
36. Marton, J. In *Lignin*; Sarkanen, K. V.; Ludwig, C.H., Eds; Wiley: New York, 1971.
37. Toth, L. *Fleischwirtsch.* **1980**, *60*, 728.
38. Domburg, G. E.; Sergeeva, V. N.; Zheibe, G. A. J. *J. Thermal Anal.* **1970**, *2*, 419.
39. Domburg, G. E.; Sergeeva, V. N.; Kalnish, A. I. In *Thermal Anal.* 3; Proc. Third. ICTA Davos; Birkhaeuser Verlag: Basel, 1971; pp 327.
40. Fenner, R. A.; Lephardt, J. O. *J. Agric. Food Chem.* **1981**, *29*, 846.

RECEIVED January 28, 1992

Chapter 19

Reaction Kinetics for the Formation of Oxygen-Containing Heterocyclic Compounds in Model Systems

J. P. Schirle-Keller and G. A. Reineccius

Department of Food Science and Nutrition, University of Minnesota, St. Paul, MN 55108

A gas chromatograph equipped with an atomic emission detector was used to collect data for kinetic studies on the formation of oxygen-containing heterocyclic volatiles formed during the heating of a model Maillard system. The model system studied contained glucose (0.5 mole) and cysteine (0.1 mole) in 400 ml water and was heated at temperatures of 80, 100, 120, and 150°C for 2 to 10 h. The formation of furfural, furfuryl alcohol, 2-acetylfuran, 5-methylfurfural, di(H)-di(OH)-6-methyl pyranone and 5-(hydroxymethyl)-2-furfural was found to follow zero order kinetics. Activation energies of these compounds ranged from 28 to 33 kcal/mole.

The Maillard reaction is probably the most important reaction occurring during the heating of foods. It is responsible for some of the most pleasant aromas (e.g. bread, roasted meat, cookies, coffee and chocolate). Unfortunately it also is responsible for some less desirable aromas both during heating and storage. The aroma of burned bread, canned vegetables and fruit as well as the staling of dry milk or dry potatoes during storage are examples of the less desirable side of the Maillard reaction and flavor. It is of interest that the same reaction, i.e. the Maillard reaction, gives us both desirable and undesirable flavors. To some extent the difference between desirable and undesirable flavor is a matter of reaction kinetics. At elevated temperatures certain reactions are favored and a desirable roasted or toasted note is formed. At room temperature, the same reaction produces compounds which elicit stale gluey notes. The better we can understand the reaction kinetics involved in the Maillard reaction, the better we can control flavor development via this reaction scheme.

A great deal of research has been devoted to the study of flavor formation via the Maillard reaction (1-3). These studies have largely involved either real foods or model systems. They most often have been qualitative in nature and empirical in design. Literally thousands of volatiles have been identified as arising from this reaction (4-6). There have been few studies, however,

0097–6156/92/0490–0244$06.00/0

© 1992 American Chemical Society

which have had the goal of developing reaction kinetics relating to flavor formation. Leahy and Reineccius (7-8) reported on the reaction kinetics of the formation of a limited number of pyrazines in model systems. The model systems contained sugar (glucose, fructose or ribose) and amino acid (lysine or asparagine). Reaction rates were also studied as a function of pH. Leahy and Reineccius (7-8) found an E_a of 27 to 45 kcal/mole.

In this study, we determined the reaction kinetics of oxygen containing heterocyclic compounds in a model system containing cysteine and glucose using a gas chromatograph equipped with an atomic emission detector (AED). While numerous oxygen-containing heterocyclic compounds were detected, the paper will focus only on those compounds which yielded kinetic data (i.e. were quantifiable in the model system when heated at three temperatures).

MATERIALS AND METHODS

Sample Preparation. Glucose (0.5 mole = 36.03 g; Sigma Chemical Co., St Louis, MO) and cysteine (0.1 mole = 4.84 g; Sigma Chemical Co., St Louis, MO) were dissolved in 400 ml of distilled water (Glenwood Spring, Minneapolis, MN) in a beaker. Temperature and pH were recorded prior to reaction. A pressure reactor (Model series 4560, 400 ml stirred vessel, monitored by Model 4842 controller, PARR Instrument Co., Moline, IL) was then filled with the entire solution, sealed and heating as well as stirring were started. The time to reach the reaction temperature was noted. (It varied between 8 min when the final temperature was 80°C and 18 min when the final temperature was 150°C.) The time at which the system reached the final temperature was the time t=0 for the experiment. During all the experiments, the stirrer was set at 60%.

At 100 and 120°C, a sample (50 ml) was withdrawn every 2 h over a period of 10 h from the vessel via a sampling port. At 150°C, sampling was done every 20 min for a period of 140 min. Nitrogen gas was added after sampling (ca. 50 ml) to restore pressure to the reactor. After cooling the sample to 25°C, the pH was measured and the sample was extracted 3 times with 5 ml aliquots of dichloromethane containing octane (10 ppm) as internal standard. The extract was then dried with anhydrous $MgSO_4$ and concentrated 300 times under a N_2 flow (only 150 times for the experiment at 150°C).

Gas Chromatography/Mass Spectrometry. A Hewlett-Packard (HP) model 5890 Series II gas chromatograph (GC) equipped with either an AED 5921A (carbon and oxygen signal) or mass selective detector MSD 5970 were used. Components were separated on a 30 m x 0.32 mm i.d. x 1 μm film thickness DB -5 fused silica capillary phase column (J&W Scientific, Folsom, CA). The temperature program used for all the experiments was: initial temperature 40°C, temperature held 1 min, temperature was then increased 5°C/min, until 250°C which was the final temperature. For both detectors, the transfer line temperature was held at 275°C. Helium was used as carrier gas in the two systems with ca. 2 ml/min flow. In the case of the AED detector, the flow of the different reagent gases were set based on the recommendations of Fox and Wylie (9). For the carbon signal (193 nm), O_2 and H_2 were used as scavenger gases and the makeup flow was 30 ml/min, while for the oxygen signal (777 nm), a mix of $H_2/N_2/CH_4$ was used with the

same makeup flow. Two μl of the flavor concentrate was injected into
the GC using a 30 to 1 split.

Compound identification was based on mass spectra (comparison to
NBS library), published retention indices and cochromatography with
authentic compounds (compounds used in kinetic study).

Analysis of data. All data obtained from the carbon channel of the
AED were normalized with respect to the internal standard. The data
from the oxygen signal were also normalized using the same factors
as were used for the carbon channel. Normalized GC areas were
plotted versus time of heating and analyzed using the kinetics
program of Labuza and Kamman (10).

RESULTS AND DISCUSSION

Comparison of the AED Carbon and Oxygen signals. The use of
selective GC detectors is generally very beneficial in flavor
research. Selective detectors can reduce sample cleanup
requirements and put less stringent demands on the chromatography
itself. It also will make compound identification easier since the
researcher will know what atoms are present in the unknown molecule.
While the chromatographer has had selective detectors available for
a number of years, the detectors available have often been less than
ideal. The sulfur detector (flame photometric) is very sensitive
but gives very poor quantitative data since it has a very limited
linear dynamic range. The nitrogen detector (NPD) is also very
sensitive but may present problems in long term stability. Both
detectors will be influenced by quenching and may give a response to
other atoms. The AED offers an ideal detector in that it is
inherently linear and highly selective. The absolute nature of its
response permits the determination of the empirical formula of an
unknown (11). Our interest in the AED for this study was to utilize
its unique ability to selectively detect oxygen-containing
compounds. While there are other oxygen detectors available on the
market, they have not been used in solving flavor problems (12).

Of the atoms which can be detected using the AED, it is least
sensitive to oxygen (9). The relative sensitivity of the carbon
channel vs the oxygen channel is evident from Fig. 1. (One should
note that the carbon channel of the AED is about 10 fold more
sensitive than the typical FID.) While the carbon channel yields a
chromatogram with acceptable area counts for reliable
quantification, the oxygen channel gives a very weak chromatogram
which lacks many compounds and is not suitable for quantification.
Thus the lack of sensitivity becomes obvious. As compound
concentration becomes less limiting (Fig. 2), the value of the AED
is clear. The AED gives a useful chromatogram that clearly
identifies molecules which contain oxygen.

The relative sensitivity of the oxygen channel to the carbon
channel can also be observed from the data presented in Figs. 4 and
5. Since these data will be discussed in the kinetics section of
this paper, they will not be discussed further here.

Flavor Formation - General Observations.

In general, compounds formed increased both in number and
quantity as either time or temperature of heating increased (see

Figs 1-5). However, some components were identified only at a
certain temperatures. For example, 2,5-hexanedione was found only at
120°C and the 1,4-hexanediene-2,3,4,5-tetramethyl was detected only
at 100°C. Apparently some compounds need a certain level of energy
to be formed but too much will result in their degradation to other
compounds and they disappear for the chromatogram. This dynamic
nature of compound formation is certainly partially responsible for
the changes in flavor that occur as one heats foods at different
temperatures.

Some compounds were detected under all conditions of the study
such as furfural, furfuryl alcohol, 2-acetylfuran, 5-methylfurfural,
di(H)-di(OH)-6-methyl pyranone and 5-(hydroxymethyl)-2-furfural
(focusing on the oxygen-containing compounds). Oxygen-containing
compounds that were present at all three temperatures were selected
for the kinetic study. We also chose to limit our discussion to
those compounds that contained oxygen in a ring structure - i.e.
furans and pyrans. While oxygen was found in other compounds, e.g.
thiophenones and thiofurans, the kinetics of these compounds will be
reported on latter in a separate publication. Therefore, only
furfural (Fal), furfuryl alcohol (Fol), 2-acetylfuran (AF), 5-
methylfurfural (MFal), di(H)-di(OH)-6-methyl pyranone (DHMP) and 5-
(hydroxymethyl)-2-furfural (HMF) were chosen for the kinetic study.

Kinetic study. This study was conducted at four temperatures (80,
100, 120, and 150), however, none of the 80°C data were useable for
the oxygen-containing compounds. Although peaks were observed in
the 80°C sample using the AED (C-193 wave length), none of the
components were identifiable since the AED is about 10,000 times
more sensitive than the mass spectrometer. We were not comfortable
using any analytical data when we could not verify peak identity and
purity via the mass spectrometer. Thus all kinetic data were based
only on three temperatures which is less than ideal.

Plots showing compound formation as a function of heating
temperature and time are shown in Figs. 3-5. Fig. 3 does not show
results from the oxygen channel because response was too low. A
visual comparison of the plots show that the data obtained from the
oxygen channel parallel those obtained from the carbon channel. It
appears that there is little difference in linearity in response
between detectors and either detector could be used for
quantification. Since the carbon channel gave a greater response and
provided data at 100°C, its output was used in kinetic calculations.

A summary of the data used to calculate reaction order is
presented in Table 1. The data obtained at 100°C fit either zero or
first order kinetics equally well. The r^2 values are essentially
equivalent except for HMF which is better fit with zero order
kinetics and DHMP which is better fit using first order kinetics.
The 120 and 150°C data are a much better fit with zero order
equations. Thus, zero order was assumed for calculating reaction
order.

A summary of the activation energy data is presented in Table 2
and plotted in Fig. 6. As was mentioned earlier, it would have been
desirable to have more than three temperatures for the calculation
of activation energy, but data were not available at 100°C so only
three temperatures were used in E_a calculations. Despite the
limited data, the E_a values should be reasonably close to actual

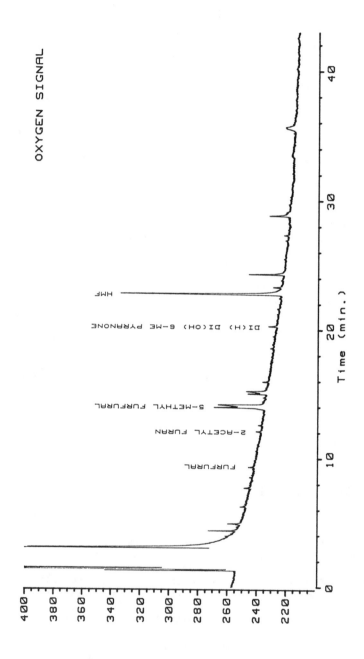

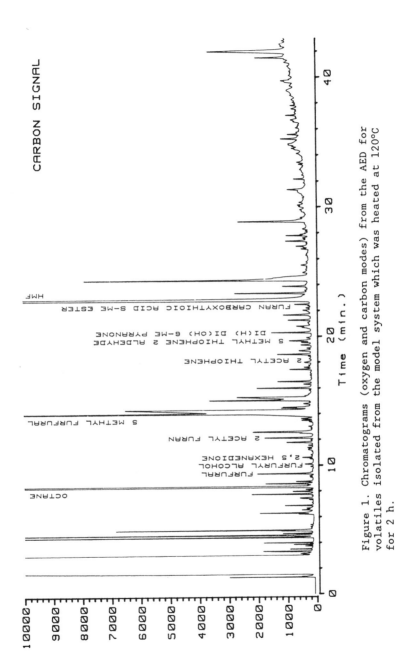

Figure 1. Chromatograms (oxygen and carbon modes) from the AED for volatiles isolated from the model system which was heated at 120°C for 2 h.

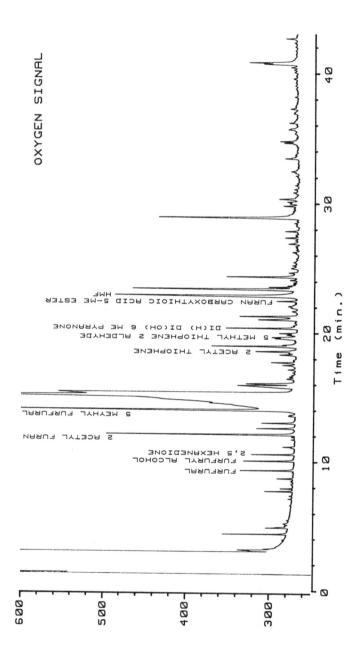

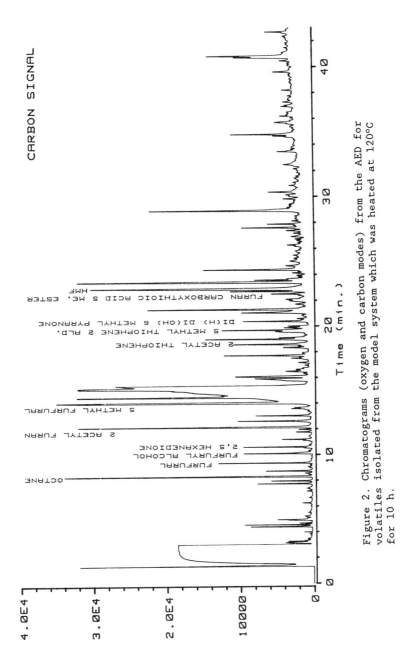

Figure 2. Chromatograms (oxygen and carbon modes) from the AED for volatiles isolated from the model system which was heated at 120°C for 10 h.

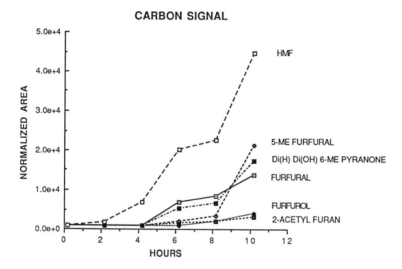

Figure 3. The influence of heating time at 100°C on the quantity of oxygen-containing heterocyclic compounds formed (data are normalized with respect to the internal standard).

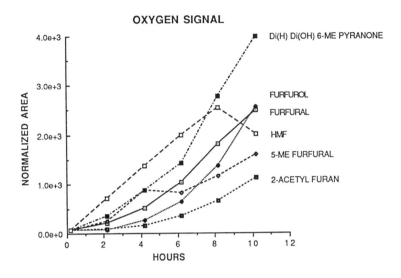

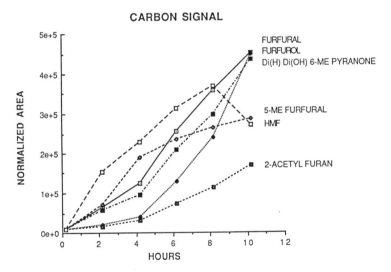

Figure 4. The influence of heating time at 120°C on the quantity of oxygen-containing heterocyclic compounds formed (data are normalized with respect to the internal standard).

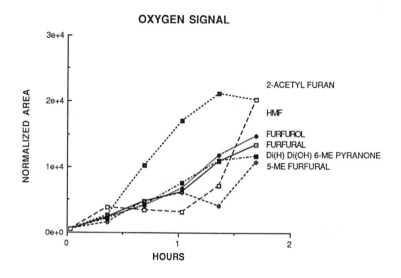

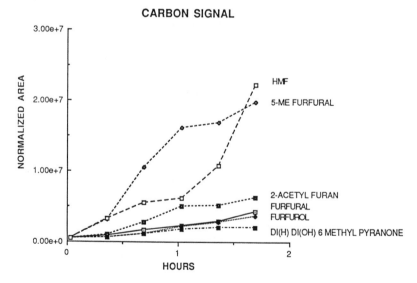

Figure 5. The influence of heating time at 150°C on the quantity
of oxygen-containing heterocyclic compounds formed (data are
normalized with respect to the internal standard).

Table 1: Determination of the reaction order

Temperature 100°C

Compound	zero order k (UNA hr^{-1})	intercept	r^2	first order k (hr^{-1})	intercept	r^2
furfural	$1.692\ 10^3$	$-4.700\ 10^3$	0.91	2.280	-9.568	0.89
furfurol						
2-acetyl furan	$4.151\ 10^3$	$-2.048\ 10^4$	0.93	2.228	-9.472	0.88
5-Me furfural	$4.866\ 10^4$	$-3.087\ 10^5$	0.80	2.469	-9.779	0.92
Di(H) Di(OH) 6-Me Pyranone	$1.377\ 10^3$	$-1.646\ 10^4$	0.73	2.236	-9.603	0.89
HMF	$4.224\ 10^4$	$-5.780\ 10^4$	0.89	1.907	-1.867	0.56

Temperature 120°C

Compound	zero order k (UNA hr^{-1})	intercept	r^2	first order k (hr^{-1})	intercept	r^2
furfural	$5.030\ 10^4$	$-5.942\ 10^3$	0.99	1.84	-0.987	0.50
furfurol	$4.217\ 10^4$	$-7.177\ 10^4$	0.84	1.902	-1.979	0.57
2-acetyl furan	$1.945\ 10^5$	$-4.551\ 10^4$	0.97	1.947	$+0.340$	0.47
5-Me furfural	$2.537\ 10^5$	$-4.792\ 10^5$	0.87	1.979	-1.006	0.54
Di(H) Di(OH) 6-Me Pyranone	$4.238\ 10^4$	$-3.634\ 10^4$	0.95	1.836	-1.107	0.54
HMF	$2.922\ 10^5$	$+6.949\ 10^5$	0.72	1.922	$+0.738$	0.45

Temperature 150°C

Compound	zero order k (UNA hr^{-1})	intercept	r^2	first order k (hr^{-1})	intercept	r^2
furfural	$6.829\ 10^5$	$-1.389\ 10^5$	0.89	11.60	-0.923	0.52
furfurol	$6.159\ 10^5$	$-1.495\ 10^5$	0.91	12.01	-1.702	0.57
2-acetyl furan	$1.165\ 10^6$	$-1.739\ 10^5$	0.96	11.96	-0.747	0.53
5-Me furfural	$3.878\ 10^6$	$-3.859\ 10^5$	0.98	12.32	$+0.119$	0.50
Di(H) Di(OH) 6-Me Pyranone	$3.258\ 10^5$	$-4.485\ 10^4$	0.97	11.41	-1.327	0.53
HMF	$3.674\ 10^6$	$-9.330\ 10^4$	0.76	12.21	-12.48	0.51

Note: UNA hr^{-1} corresponds at: Units of Normalized Area / hour

Table 2: Activation energy for selected flavor compounds
            assuming zero order reactions
                Ln (k) vs 1/T (K^{-1})

Compound	slope	intercept	r^2	E_a (kcal/mole)
furfural	$-1.77 \ 10^4$	$5.49 \ 10^1$	0.94	35.18
2-acetyl furan	$-1.82 \ 10^4$	$5.68 \ 10^1$	0.87	36.20
5-Me furfural	$-1.86 \ 10^4$	$5.93 \ 10^1$	0.86	37.03
Di(H)-Di(OH)-6-Me pyranone	$-1.54 \ 10^4$	$4.91 \ 10^1$	0.85	30.73
HMF	$-1.41 \ 10^4$	$4.82 \ 10^1$	0.83	28.14

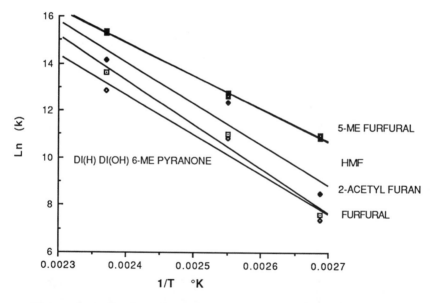

Figure 6. Arrhenius plot of zero order kinetic data for the
formation of oxygen-containing heterocyclic compounds.

values since there were no discrepancies in the data that would introduce substantial error or uncertainty.

The activation energies found for oxygen-containing heterocyclics ranged from 28 to 37 kcal/mole. These values are quite close to what Leahy and Reineccius (7-8) reported for the formation of pyrazines (27 to 45). While E_a for the oxygen-containing compounds studied here may be a little lower than the pyrazines, more temperatures would have to be used to obtain more accurate values to determine if significant difference exist.

CONCLUSIONS

In this study we have presented data to demonstrate the utility of operating an AED in the oxygen mode for flavor studies. While sensitivity in the oxygen mode was not equivalent to the carbon mode, it was valuable in compound identification and provided quantitative data free from interfering compounds (i.e. nonoxygen-containing compounds).

Generally speaking, the quantity of the flavor components produced during heating of model systems increased with time and temperature of heating. The formation of oxygen-containing heterocyclic compounds was found to follow zero order reaction kinetics. Activation energies (28 - 37 kcal/mole) were similar to those reported in the literature for pyrazine formation (27 - 45 kcal/mole) by Leahy and Reineccius (7-8).

LITERATURE CITED

1. Bessiere, Y.; Thomas, A.F. *Flavour Science and Technology*. J. Wiley & Sons: New York, 1990.
2. Finot, P.A.; Aeschbacher, H.U.; Hurrel, R.F.; Liardon, R. *The Maillard Reaction in Food Processing, Human Nutrition and Physiology*. Birkhauser Verlag: Switzerland, 1990.
3. *Thermal Generation of Aromas*; Parliment, T.H.; McGorin, R.J.; Ho, C.T., Eds.; American Chemical Society Symposium Series 409; American Chemical Society: Washington, 1989.
4. Maarse, H.; Visscher, C.A. *Volatiles Compounds in Food*; TNO-CIVO: Netherlands, 1989, Vol. 1.
5. Maarse, H.; Visscher, C.A. *Volatiles Compounds in Food*; TNO-CIVO: Netherlands, 1989, Vol. 2.
6. Maarse, H.; Visscher, C.A. *Volatiles Compounds in Food*; TNO-CIVO: Netherlands, 1989, Vol. 3.
7. Leahy, M.M.; Reineccius, G.A. In *Flavor Chemistry*; Teranishi, R., Buttery, R., Shahidi, F. Eds.; Advances in Chemistry Series No. 388; American Chemical Society: Washington, 1989, pp 76-91.
8. Leahy, M.M.; Reineccius, G.A. In *Thermal Generation of Aromas*. American Chemical Symposium Series No. 409; Parliment, T.H., McGorin, R.J., Ho, C.T. Ed.; American Chemical Society: Washington; 1989, pp 196-208.
9. Fox, L.; Wylie, P. *Hewlett-Packard Appl. Note*. 1989, 228-75, pp 1-3.

10. Labuza, T.; Kamman, J. F.   In *Computer Aided Techniques in Food Technology*; Saugy, I., Eds.; Marcel Dekker: New York Inc. 1983, pp. 71-113.
11. Wylie, P.L.; Sullivan, J.J.; Quimby, B.D.   *J. High Resol. Chrom.* **1990**, Vol 13, pp. 499-506.
12. Steinmuller, D.   *Am. Lab.*   **1989**, *Vol.* 120, pp. 122-125.

RECEIVED January 28, 1991

INDEX

Author Index

Affiliation Index

Subject Index

Production: Margaret J. Brown
Indexing: Deborah H. Steiner
Acquisition: Barbara C. Tansill
Cover design: Peggy Corrigan

Printed and bound by Maple Press, York, PA